UNBELIEVABLE

ALSO BY STACY HORN

THE RESTLESS SLEEP:
INSIDE NEW YORK CITY'S COLD CASE SQUAD

WAITING FOR MY CATS TO DIE:
A MORBID MEMOIR

CYBERVILLE: CLICKS, CULTURE, AND
THE CREATION OF AN ONLINE TOWN

INVESTIGATIONS INTO

GHOSTS,

POLTERGEISTS,

TELEPATHY,

AND OTHER UNSEEN PHENOMENA

FROM THE DUKE

PARAPSYCHOLOGY LABORATORY

UNBELIEVABLE

STACY HORN

An Imprint of HarperCollinsPublishers

HarperCollins books may be purchased for educational, business, or sales promotional use. For information, please write: Special Markets Department, HarperCollins Publishers, 10 East 53rd Street, New York, NY 10022.

A hardcover edition of this book was published in 2009 by Ecco, an imprint of HarperCollins Publishers.

P.S.™ is a trademark of HarperCollins Publishers.

FIRST PAPERBACK EDITION PUBLISHED 2010

Designed by Jessica Shatan Heslin / Studio Shatan, Inc.

Library of Congress Cataloging-in-Publication Data is available upon request.

ISBN: 978-0-06-111690-2

10 11 12 13 14 ID/RRD   10 9 8 7 6 5 4 3 2 1

To the men and women of the former
Parapsychology Laboratory
of Duke University

What we are dealing with is a vast and half-lit area, where nothing seems believable but everything is possible.

—CARL JUNG

UNBELIEVABLE

PREFACE

Every ghost story begins with a love story, and usually more than one. Once they are untangled, you will always find eternal love, unbearable loss, and unconquerable fear.

Everywhere, every minute, people all over the world are desperately begging God and any other power they can think of to not take someone they love, their child, their husband or wife, mother or father or friend. And finally, at the end, don't take me.

There is no spot on earth that is free from loss. On this street, or in this room, someone lay down or was put down and was no more. Someone held someone else for the last time here. Rivers and lakes and oceans are full of people who vanished beneath the surface and were never seen again. Wherever you are standing, wherever you call home, someone left the earth there. Everyone we love dies and disappears.

Something more substantial than a memory must survive of all that love. It's unthinkable that the dead are truly and completely gone. And if the dead are not completely gone, we, as every generation that came before, are compelled to look for whatever remains.

What is death but the end of all we love? Ghosts are what

survive of love. Real or unreal, they are a testament to love, and the hope that no matter what, love lasts.

The men and women of the Duke Parapsychology Laboratory were scientists. They never would have phrased it this way. But when all is said and done, as they tried to prove that death is not the end, what they were really trying to prove is that love lasts forever. The problem was how to scientifically demonstrate that life and all the feelings that go with it survive death. A medium relaying messages of continuing love from a dead wife might be enough for an inconsolable widower, but it would never be enough for the scientific community, which demanded not only more convincing evidence but also experiments that could be reliably repeated to produce consistent results. To move an idea out of the realm of belief and into the world of accepted fact, others must be able to verify your results. There are no shortcuts to this process, and no exceptions. Like those we pray to when death is imminent, the scientific method is immune to longing, hope, and pleas.

■

I've always been drawn to the dead and forgotten. My last book, about the New York Police Department's Cold Case Squad, had me immersed in unsolved murders for years. My plan was to find people who had been murdered and denied justice, and for whom it was over in every possible way. I was going to bring them back to life, if only on a paper stage. When the NYPD allowed me into warehouses and basements and closets to look inside hundreds of boxes of old homicide cases, many unopened for decades, I was practically hypnotized; it had the excitement of a treasure hunt and the hope of resurrection. The problem was that, while the research couldn't have been more fascinating, it was all ultimately heartbreaking. Everything I did came back to someone who had died, and always in some horrible and unthinkable way.

By the time I was done, I was ready to tell my agent that my next project would be a coffee-table book entitled *Puppies of North America*.

But then I thought: What's more fun than a ghost story? I've never been able to resist the paranormal. So I started looking for a story that was supernatural and scary and still sad, but without all the bloodshed. I began with a haunted house in New York City that I'd heard about when I was a kid and had never forgotten. This led to a group of scientists at Duke University who specialized in parapsychology.

With few exceptions, scientists have always been disdainful of accounts of the paranormal and the idea of studying it rigorously. But there was a moment when science discovered a kaleidoscope of invisible forces, waves, and particles underneath a now thinner veil of reality, and that the universe is expanding. At that moment there was just enough give in the scientific community that phenomena that had always been dismissed as paranormal could be approached differently. In this brief window of opportunity, Duke University opened the Parapsychology Laboratory. I'd heard of the lab over the years, but when I went down to North Carolina I had little idea of what went on there.

It was the Cold Case Squad all over again. There are over seven hundred boxes of lab archives at Duke University, and what I found inside was the story of scientists who were like paranormal detectives, but without the horrible murders, although in the end there were some of those, too. Like the cold case detectives who worked on homicide cases no one else could solve, the scientists at the Parapsychology Lab were looking at "cases" other scientists either didn't want to or couldn't explain. I was once again completely absorbed in material that few had looked at in decades. I found death and loss wrapped up in every box I opened, but this time, behind it all, was the hope of something more than a metaphorical resurrection.

The lab was headed up by a man named J. B. Rhine. Outside of parapsychology circles, Rhine is not well-remembered today, but in his time, he was the Einstein of the paranormal. Between 1930 and 1980 the media seized on Rhine and reported on nearly every bit of groundbreaking research that came out of the lab. From *Life* magazine to *Reader's Digest* to the *New York Times* and the *New Yorker*, newspapers and magazines were eager to pass on any shred of evidence of life after death to a public that anticipated paranormal breakthroughs to rival those being heralded in other areas of science on what felt like an almost daily basis. In turn, the public would bring thousands of supernatural events to the lab's attention. The scientists at the lab were flooded with mail and visitors, legendary and otherwise, including Albert Einstein, Upton Sinclair, B. F. Skinner, Richard Nixon, Aldous Huxley, Arthur Koestler, Carl Jung, Bill Wilson (the co-founder of Alcoholics Anonymous), and Jackie Gleason. Timothy Leary flew down to Duke in the sixties and got Rhine and some other scientists there to take their first psilocybin trip to see if this extraordinary drug could enhance the remarkable mental abilities they were trying to understand—ESP, psychokinesis, and clairvoyance. Helen Keller put her fingers to Rhine's lips when they met and told him that she experienced ESP often. Rod Sterling, who really wanted to believe in the supernatural and ESP, kept an eye on their progress. The Army and the Navy gave the lab contracts to study extrasensory perception in animals, and the Air Force built an "ESP machine," basically a simple computer that automated parts of the experiment. Defense industry contractors like General Dynamics stopped by at the request of the government's Advanced Research Projects Agency to see what the lab was up to. When word came from behind the Iron Curtain that the Russians had a parapsychology lab too, there was talk of a Manhattan Project for the paranormal. The CIA would eventually spend millions exploring mind control.

Industry also took an interest in the lab. As early as 1938, Rhine began talking to IBM about building an ESP machine. Like the Air Force later, Rhine wanted to see how technology might help increase their understanding of ESP, and at the start IBM was willing. Representatives from other companies like AT&T, the Zenith Corporation, and Westinghouse got in touch with Rhine in order to conduct experiments to learn more about these mental powers that might one day have a real-world application. Practical businessmen and hard scientists like Alfred P. Sloan, of Sloan-Kettering, and Chester F. Carlson, founder of the Xerox process, were so excited by Rhine's results that they invested in the lab's experiments out of their own pockets just to see what else he could find out. The very prestigious and apparently intellectually daring Rockefeller Foundation contributed to the laboratory at the same time they were funding another academic pioneer, Alfred Kinsey.

Rhine and Kinsey had a lot in common. Kinsey was an entomologist who went on to pioneer the scientific study of human sexuality, and Rhine started out with a PhD in botany. Imposing, driven men, they were both a complicated mix of strengths and weaknesses. Rhine, like Kinsey, had all the right qualities needed to lead research in an area almost entirely unexplored by academia. Tall, good-looking, and seductive, Rhine had the allure of an intellectual rebel. He was always able to gather fiercely loyal students, staff, and associates to the parapsychological cause. The famed anthropologist Margaret Mead once stood up to the most formidable group of scientists in America to fight for his work. But also like Kinsey, Rhine possessed some equally impressive flaws. An ex-marine, Rhine loved a fight, which was useful. Convincing the skeptical, frequently outright hostile scientists to accept the results of the lab's experiments was going to take a battle. But Rhine loved battle a little too much. He didn't always know when to exit the field, and sometimes saw enemies where his own soldiers stood. To this day there are people whose pain and anger remain so acute they can barely speak about him. But there are also women whose hearts still flutter at the sound of his voice on tapes and men who are willing to stake their reputations on his integrity. For his entire life, Rhine was both celebrated and ostracized. He'd lecture to standing-room-only crowds, from Lockheed Missiles & Space Company on the West Coast to riveted academics at Harvard University on the East, but he sometimes had to respond afterward to ceaseless—and redundant—challenges of fraud, delusion, and incompetence.

Although most of the lab's work was with ESP, which lent itself to experimentation most easily, the seven hundred boxes that made up the lab archives contained some fabulous surprises. One day I came across a pile of letters from a distraught priest in Maryland. His letters described unusual events surrounding a child in a home outside Washington, D.C. It sounded exactly like the movie *The Exorcist*, right down to the projectile vomiting. And, in fact, it was the true story on which the author William Peter Blatty had based his bestselling book. The priest told Rhine he thought the disturbing events might be the work of a poltergeist and asked if the lab would be interested in investigating it—they would and they did, to a point.

Rhine always did his best to keep an open mind about anything supernatural, but what he couldn't test in the lab was not going to help his search for irrefutable proof of life after death. For example, he would later send members of his staff to New York to investigate a poltergeist case so dramatic and perplexing the local police opened an investigation that remains unsolved to this day. Neither the detective assigned to the case nor any of the various experts he consulted could explain how a statue was able to float above a table unaided, make a tremendous explosive sound, and then fall to the floor in one piece. Rhine was always very careful about what he deemed paranormal, as he had a scientific audience to answer to, but while Rhine and his colleagues would hesitate to claim any of the sixty-seven events in this case paranormal, they did conclude that many of them could not be explained by normal means. Many famous mediums and psychics of the day refused to set foot in the lab due to Rhine's prudence. A lucrative career could be destroyed by failing to convince the dispassionately observing scientists whose opinions would immediately be picked up by newspapers across the country and throughout the world.

Disappointments and uncertain conclusions helped keep Rhine in the lab, and his focus was on extrasensory perception, the one area in all the supernatural territory in which he felt that science and belief could meet, be tested, and begin to be understood. The lab eventually made discoveries about telepathy that are being rediscovered today. The United States government would later spend millions on an aspect of ESP that Rhine's wife discerned in the fifties. Books and papers about audio hallucinations published today detail discoveries similar to those made by the lab in their ongoing survey of ghost stories. Right now the study of consciousness is one of the hottest fields in science, and Rhine always believed that consciousness was the key to ESP.

But while Rhine tried to focus on consciousness and ESP, the aspect of the paranormal that looked the most like science, his supporters and many of his colleagues were driven by an interest in life after death to leave the lab and head into the field where faith and wishful thinking, rather than hard data, made a case. Rhine would spend a lifetime disappointing these people. In the beginning, I was among them. When Rhine didn't follow up on some strange but fascinating area of the paranormal, like near-death experiences, I thought, "Come on. Not worth even a look? *Really?*" But then, very quickly, I became more excited by the possibility that Rhine and his colleagues may have found empirical proof that one of the phenomena they were studying—ESP—was genuine.

The ultimate, heartbreaking fact is that if anything was being missed in all this research and all this work, it was not the steps Rhine didn't take to answer the question of life after death, but just how close he got. When I began my research I was somewhere between the faithful and the scientific. I didn't believe in ghosts when I went down to North Carolina to study the work of the Parapsychology Laboratory, and I didn't believe in ESP either. I'd seen the movie *Carrie* when I was in college, and I tried to move a pencil across my desk with my mind and failed. But I'm human, and who doesn't want to believe that the mind is not as limited as we thought and death is not the end? A ghost story is great, but scientific evidence that there really is something paranormal in the world is better. I knew that no one has definitely proven that there is life after death, but Rhine and his colleagues conducted lots of experiments at Duke. What if one of them worked?

What I discovered, among many other things, is that one of them did.

LETTERS RECEIVED
AT THE DUKE
PARAPSYCHOLOGY
LABORATORY

I do not know whether or not you answer or act as consultant for an individual 'problem,' but as I know of no one else who might understand, explain (or dismiss) a particular concern, may I presume upon your time?

. . . this is what occurs . . . 'Someone' stands just behind my right shoulder . . . I know that he is there, though he does not speak. He is dark, gaunt, shadowed . . . he may be hungry . . . I feel strangely comfortable in his presence. I feel as though he needs me . . . Can this be explained?" 1960.

"From birth this little girl now aged twenty-nine months, was different from other infants . . . At one year she was already imitating the family canary, dog, cat and a lamb, cow, frog and chickens . . . but at 19 months she began to prognosticate with an accuracy most startling. . . Floods and earthquakes. Plane crashes. The change in Superior Court. Roosevelt and his illness. She has made predictions which I fear to write down here, lest I should be suspected in some way with the results . . . May I have an answer soon?" 1938.

"On May 5th at 4:40 p.m. Tony Bradford . . . left the White Rock Airport in Dallas, Texas . . . At Houston, he picked up his fiance . . . They left Houston at approximately 8:45 p.m. About fifty miles north of Houston, a line of thunderstorms apparently caused their plane to crash. As of this writing no trace of the plane has been found . . . Would you please put this problem before the people having the ability to judge the exact location of the airplane . . . The search has been given up since the 20th of May." 1961.

"During the past months, the Boston area has been plagued with a strangular [sic]. Ten women have been strangled, and the police have no concrete clues. I mentioned the work in ESP to the Homicide Department. The detective with whom I talked said the authorities had thought of ESP work in dealing with crime but were skeptical. However, they would be interested in more information." 1963.

"I am in jail at Asheville, N.C. waiting on trial for first degree murder. I know almost to the day when I will be dying so I think I should tell someone of your calibar [sic] so the research can be furthered . . . I have been able to foretell what is going to happen in the future . . . I even saw the face of the man of [sic] whom I am accused of killing and even knew his first name but up until the time of his death I had never seen or heard anything about him . . . I will be going to court the last day of August so I hope you will contact me very shortly either in person or by letter." 1964.

[The writer was sentenced to life the following month.]

"I understand that there were some fragments of a flying saucer sent to Duke for analysis that were at one time actually put on display there and were handled by a number of people but were unanalyzable because they could not be broken down into their constituent parts. Do you know of this incident . . . I have been investigating the 'saucers' since 1946 . . ." 1957.

"August 16, 1959, I contracted polio which left me completely paralyzed and unable to breathe without the aid of respiratory equipment. My physical condition has made it mandatory to utilize the one thing I have left—my mind. I have done extensive reading on hypnotism, ESP, mind reading, clairvoyance, spiritualism . . . If you could recommend some good books along this line it would be greatly appreciated." 1961.

"Help me please . . . I'm feeling I'm breaking at the edges and for my child's sake I'd like to get this problem resolved . . . I have a loved one buried that I very much want you to examine the body as I think he returned to life after burial and is being held alive by electric current in some tranced state. Even if you think I'm wrong help me to *know* I am wrong." 1959.

"My wife together with me purchased an orange squeezer hand operated—poorly constructed and inexpensive. She bought it for a girl friend of hers, who (in my opinion) was <u>not</u> a really good friend. In my own mind, I wished & concentrated that when she operated this thing, the handle would break and cut her hand (not a nice thing to do, I admit now). Lo & behold <u>THAT</u> is <u>exactly</u> what happened to her. She required stitching by a doctor. I would be <u>most</u> grateful to you to hear from you and get <u>YOUR</u> reaction." 1958.

"Two weeks ago I received word that my husband, an army doctor, had been killed in Germany. Since then I have read several books on psychical research, including your New Frontiers of the Mind . . . I wish to write . . . asking about mediums of good character . . . My husband and I had been married only fourteen months, but we felt somehow as if we had known each other always. It seems, even now, preposterous that he will not return and that we shall not go on with a life where every obstacle existed only for the joy of overcoming it together." 1945.

"I am the owner of what appears to be a poltergeist haunted house . . . one that is rented to tenants . . . they have been complaining of . . . bed clothing pulled off, knocking on the walls, cold drafts thro' the floor . . . the sister once saw an apparition of a man dressed in something like an intern's suit, standing in the room—then disappearing . . . a former tenant also complained of odd experiences . . . would you be so kind as to advise me if you know to whom in Southern California I might appeal for a reference to a really honest and genuine medium?" 1961.

"You received a letter last year . . . sealed with cellophane. On it was a warning for the one who should read it to possess a strong will. Asking you to perform an experiment to test the distance of my ability to communicate mentally with people . . . My girlfriend received a nervous breakdown from this. I am sure a few words from you will help a great deal in convincing these doctors that there is such a thing as mental telepathy . . . This place is the Psychological Institute and Hospital of New York State." 1940.

". . . on one of those warm moonlight nights I woke up with a terrific start . . . I heard someone walking in the kitchen . . . Imagine my shock when a small man wearing a striped bathrobe and something like a turban wrapped around his head peered seemingly curiously at me . . . The figure withdrew and I waited for the footsteps, my eyes riveted on the black space in the door frame . . . Once more the little man appeared, this time only his head, staring at me . . . Looking forward to your reply with much interest . . ." 1958.

ONE

Before a small, unknown Methodist college was transformed into Duke University in the late 1920s, the city of Durham had been a backwater, known mostly for minor league baseball and cigarettes. The outskirts were bleak, and the sickly sweet smell of tobacco leaves being cured in warehouses was everywhere, wafting down the streets and on past the run-down shacks that housed the factory workers. Duke University's campus, however, was a lush, green, magnolia-scented paradise. An infusion of millions of tobacco dollars had created a rebirth, coinciding with the arrival of J. B. and Louisa Rhine, who'd been searching the country for just such an academic haven. Duke was the clean slate where they could begin their scientific pursuit of the paranormal away from the darkened parlors of charlatan mediums and under the guidance of an enthusiastic and protective mentor, Dr. William McDougall, the head of Duke's nascent psychology department. In the midst of all the building and revitalization that surrounded them, J. B. Rhine and Dr. McDougall talked mostly of death.

Rhine had spent the past six months studying the desperate and repeated attempts of a public school administrator named John Thomas to communicate with his recently deceased wife through mediums. The results were thrilling—the mediums were frequently able to provide facts known only to Thomas. They inspired a suspension of Rhine's hypercritical disbelief, but after all his efforts, Rhine was at an impasse. He'd done everything he could to verify the facts contained in the mediums' messages, even traveling hundreds of miles to a small centuries-old cemetery in upstate New York to confirm a few obscure séance-transmitted details. The information the mediums were communicating was almost always correct. But the number of verifiable facts, however staggering, did not prove that the mediums were getting their information from beyond the grave. While he had eliminated the possibility of fraud or a few good guesses, Rhine knew there could be another explanation. Telepathy is the ability of one mind to communicate directly with another without the use of any of the known senses. The mediums could have gotten their facts from the man's dead wife, but they could also have gotten them from the mind of someone living, like the husband himself.

So which was it, dead wife or telepathy? There was no way to answer the question scientifically—Rhine obviously couldn't get the dead wife into a laboratory, and while there had been some work toward substantiating telepathy in the past, the current "status of experimental telepathy" was nowhere near the point of explaining the mediums' results. But McDougall disagreed. The evidence for telepathy was "astonishingly good," he insisted. The problem was, a consistently reliable experiment under controlled settings hadn't been found. The two men decided that before he did anything else, Rhine had to design just such an experiment. McDougall, it turned out, was in a unique position to help make that happen. The study of the paranormal and life after death was as much a passion of his as Rhine's, and he'd left a prestigious position at Harvard for the promise of funding and support for psychical research from the president of Duke University himself. McDougall would do everything in his power to further Rhine's search for proof of life beyond death. Before a decade was up Durham was no longer just a tobacco town. By 1934 Duke University and Rhine's amazing experiments in what would come to be known as ESP, extrasensory perception, would captivate a nation. "The success of Rhine's E.S.P. work," McDougall proclaimed, "is about the only bright spot in a dark world." It was the first hard evidence that the elusive proof of life after death might be out there.

The journey to Duke had started in a dimly lit parlor in Boston, Massachusetts, with a beautiful woman who claimed she could talk to the dead. It was 1926, and mediums were all the rage, but Boston's Mina Crandon, the wife of the respected surgeon Le Roi G. Crandon, had every medium in America beat. In séances conducted in a fourth-floor parlor in fashionable, relentlessly respectable Beacon Hill, the captivating Mina, who went by the pseudonym Margery, held out this promise: You will never die, never disappear forever, and now, come close and hold my hands, and my feet, just to reassure yourself that this is all real, it's not a trick, and death is not the end. Mina often wore only slippers, stockings, and a dressing gown, which her husband would part so her guests could stare, in the name of science and discovery, to see the ectoplasmic emissions that sometimes came from her mouth, her ear, and from between her legs. Sex and immortality. What could be more irresistible? Some didn't know how to react. But then Dr. Crandon would invite them to reach out and touch her. At very special sittings, held in near total darkness, Crandon would flash a red light, and from between the spread legs of a now naked Mina would emerge what was described as a frightening "flaccid," "tongue-like projection." The Crandons proclaimed it the ectoplasmic emission of the hand of Mina's brother Walter, who had died in a railroad accident in 1911. It was horrifying. According to an account written by Thomas R. Tietze, one witness said the end "was broken up irregularly like amputated fingers." But no one looked away. Do you want to touch it, Dr. Crandon would ask. Many recoiled, although one sitter said it felt like a woman's breast. Others said it felt like cold, raw beef or wet rubber. These were very effluent times, and the paranormal was frequently wet, dripping, and organic. Other mediums would expel goose fat–lubricated cheesecloth. Harvard professors later concluded that Mina's "hand" was constructed from the lung tissue of an animal. Professor McDougall, then still at Harvard, was the first to state the obvious. "The more interesting question is—How did it come to be within the anatomy?" Mina never permitted the kind of inspection needed to answer the question, but she did once ask Hereward Carrington, a researcher on the committee from *Scientific American*, "Wouldn't you like to kiss me?" The magazine had offered twenty-five hundred dollars to anyone who could prove they had psychical abilities, and Carrington was there to investigate her claims. "What was I to do? She was there in my arms," Carrington later pleaded for understanding. "She was making advances to every man in sight," another visitor confirmed. Carrington would sit with Mina forty times.

During these sittings Mina would rest her feet on his knees and hold his hands. Carrington was the only one on that particular committee, which included Harry Houdini and avoid Mc-Dougall, to vote in her favor. Ultimately, the *Scientific American* committee remained neutral, neither pronouncing her genuine nor fake, but their investigation created a controversy that only made Mina more famous than ever. Other groups would insist she was authentic and defend her forcefully and passionately. Sir Arthur Conan Doyle even presented Mina with a silver cup that read, "in recognition of your heroic struggle for truth."

And so with her husband in the same room, men continued to flock to the Boston parlor to gaze at Mina's breasts and vagina, and bind her, touch her, kiss her, and make her smile, all with Dr. Crandon's approval, and all the while pretending that there was absolutely no sexual context.

Into that overheated, erotically charged townhouse walked Dr. Joseph Banks Rhine. Known as J.B., and Banks to his friends, the thirty-year-old Rhine was accompanied by his wife, Dr. Louisa Rhine, whom everyone called Louie. Rhine had started out intending to become a minister, but knowledge had replaced faith while he was in college, and he renounced religion. Since academia didn't yet have an answer about life after death, Rhine started looking for other ways to approach the problem. J.B. and Louie had heard about spirit photography and other popular vehicles to reach beyond the grave. They listened to the experts of the day, such as Arthur Conan Doyle, who was absolutely certain that he had communicated with his dead son. His conviction was impressive, and his suggestion that psychic phenomena could be subjected to scientific investigation reinforced the Rhines' growing interest. The work Rhine had been pursuing in botany and biology no longer seemed real or vital enough to fill an entire lifetime.

By the end of the year J.B. and Louie had resigned their positions at West Virginia University, sold all their furniture, and come to Boston, where some of the most exciting work in psychical research, as it was then called, was being conducted, and to Harvard University and the distinguished British psychologist Dr. William McDougall, who, like them, believed psychical research was a suitable subject for university study.

The Rhines first sat with a medium named Mrs. Minnie Soule, but the results were disappointing, and so they turned to the famous Mina Crandon, who had impressed so many of their new Boston colleagues. When the Rhines climbed the stairs of the Crandons' townhouse at 10 Lime Street on a Friday evening in midsummer, July 1, 1926, their expectations were enormous.

Things went wrong from the start. The Rhines were told that they were going to get the "standard" sitting, which was meant for newcomers and not scientists. The standard sitting began with dinner and drinks—adding alcohol to the already heady mix of sex and the promise of life everlasting. On that particular night, Dr. Crandon offered a glass of champagne to teetotaler Rhine, who declined. When the meal was completed, Mrs. Crandon left the room to change into her dressing gown and slippers.

The Rhines were not allowed to look around the parlor before the séance began, but they were given the various contraptions that would be used later in the evening to examine, like the "voice machine." This was basically a tube and a mouthpiece that was inserted into Mina's mouth, making it impossible for her to speak and thereby proving it was the dead talking and not Mina. The Rhines also examined Mina's trumpet, something common to mediums of the day. Trumpets were used to project the voices of the dead and were essentially glorified megaphones. The Rhines looked everything over and could find no obvious trickery in any of Mina's devices.

Mina entered the room. Wires were attached to her wrists and ankles as she sat in a chair in a cabinet, her hands coming through the sides, resting on shelves. A leather collar was fixed around her neck and fastened to the back wall. Once again, if they'd been asked, everyone at the séance would have insisted that these measures were taken to prove that it was not fraud. (It had nothing to do with seeing a beautiful woman in bondage.) The front door to Mina's cabinet was open, and the guests formed a semicircle around her. Rhine sat at her immediate right.

The lights went out. From then on they would sit in darkness except when Dr. Crandon turned on his red light so they could see a particular effect. When Mina fell into a trance, the spirit of her brother Walter assumed command. Walter would keep up a friendly banter throughout the trance portion of the evening.

Rhine would later describe the séance as a performance with seven acts. The details seem quaint now. A basket floated above Mina's head in one act. In another, the guests were asked to volunteer personal items that Mina would identify in the dark. The voice machine was brought out, and once the mouthpiece was positioned, Walter recited the Lord's Prayer in German. Mina's trumpet appeared and was heard bumping around inside of the cabinet for a bit before it was held up by unseen forces and thrown into the room. At one point Rhine saw Mina use her foot to kick the trumpet closer when it had temporarily gotten away from her. There was also so much play in the wires that tied Mina's hands that Rhine could hear the tug of the wire whenever "Walter" was said to have thrown something.

Two weeks later Rhine wrote to the American Society of

Psychical Research's board of trustees. The Rhines were members, and the society's support of Mina was part of the reason for their initial enthusiasm. "I am disgusted, not only with the case but with the attitude our *Journal* has taken on it, sponsoring it before the scientific world. The whole case is sure to crash some of these days and where will our reputations be then? We will be the laughing stock of the world for years to come."

J.B. and Louie submitted a scathing account of the sitting to the society's journal. The society sat on it a little too long for the Rhines, and so they decided to try the *Journal of Abnormal and Social Psychology* at the same time. Although simultaneous submissions are considered bad form, Rhine wanted to secure a professional reputation as a serious scientist, and the American Society of Psychical Research was beginning to lose its reputation for doing serious scientific research. Their report was accepted by the *Journal of Abnormal and Social Psychology*, Rhine resigned his membership in the American Society of Psychical Research, and in the beginning of 1927 their paper was published.

Their report described the event in damning detail. Where the Rhines couldn't exactly explain Mina's deception they offered, "The sister art of magic offers ready-made so many instruments and methods that it is foolish for us to ponder long over the question of 'just how she did it.'" And then the Rhines make an argument that would be turned against them for the rest of their lives: "If we can never know to a relative certainty that there was no trickery possible, no inconsistencies present, and no normal action occurring, we can never have a science and never really know anything about psychic phenomena."

There is a note of cruelty at the end of the Rhines' report. Mina did this, they wrote, to create a bond between herself and her husband, who had been married twice already and who might one day start looking for the next Mrs. Crandon. Mina knew her husband had a morbid fear of death and an intense interest in psychic affairs, and this was her way of holding on to him. In an academic journal the Rhines' allegations seem jarring and out of place—it reads like gossip and is far from the sober, scientific tone of the rest of the report. Still, this was J. B. Rhine's first salvo in a lifelong battle to bring the paranormal out from the darkness of the sexually infused parlor and into the bright and unemotional light of the laboratory—and he pulled no punches.

Things were never the same after that for Mina. Later, in 1939, what Mina had feared the most happened: Her husband died and she was alone. After that, she gained weight, lost her looks, and started drinking. She was so depressed she once tried to throw herself off the roof at Lime Street in the middle of a séance. "I attended about 5 nights during the week," a friend of Rhine's wrote to him, "and Mina was drunk every night . . . Only on one occasion (Saturday morning) was she sober." Poor Mina. She was never out to exploit anyone's grief and she always refused to contact dead relatives. She was just a scared, uneducated farm girl doing the only thing she could think of to survive. The boozy and tragic sideshow finally came to an end when Mina died of cirrhosis of the liver in 1941. She was only fifty-two.

The Rhines' report was the beginning of Mina's downfall, but for others it became a call to arms. There's a story that's often repeated that just after the report came out, Bostonians opened their morning papers to this simple but blunt advertisement, which had been placed by Sir Arthur Conan Doyle: "J.B. Rhine is an ass." When various newspapers broke the story of his skeptical report about Mina, Doyle and others rushed to her defense. Members of the two premier psychical research organizations at the time, the Boston Society of Psychic Research and the American Society for Psychical Research, shot off angry letters to the editors calling Rhine a knave and a fool and at best, immodest.

While there was something exciting about taking on the paranormal establishment, the winter of 1927 was not a good time for J. B. Rhine. Just as their account became public his mother died unexpectedly. She went fast, and for that Rhine was grateful, but it also made him desperate for comfort. "I must find a place for her, yes, must by all the laws or ways of thought. That, or turn my back upon her memories and forget her." The question about life after death was even more important now, and Rhine was ready to battle the overly credulous Conan Doyles of the world to come up with a real and testable way to find the answer.

The same month the Rhines sat with Mina in Boston, the man who would bring J.B. and Louie to Duke was having an entirely different encounter with a medium. John Thomas had lost his wife, Ethel, two months before, when she'd gone into the hospital for what was supposed to be a simple operation but instead had died on the table. Thomas had such a hard time accepting her death that he began looking for her everywhere. The spirit, the essence of whom he loved, he believed, couldn't just be gone. If she still existed somewhere in some way, then there was the possibility that he could reach her. Thomas, who directed the finances of all the public schools in Detroit, was a scientific man, and sitting with mediums was both an act of desperation and a serious investigation.

Thomas had better luck than the Rhines. When he sat with

Minnie Soule, the first medium to disappoint the Rhines, the results were stunning. She didn't just get a few details right about his deceased wife, she got sixty-two things right. Thomas kept meticulous notes of the sittings and the information the mediums communicated. Unlike the dark séances of Mina and others, the Soule séances were conducted in the light and were markedly more professional. What's more, John Thomas recognized his wife's personality in the communications. The jokes that were made were very much Ethel's particular sense of humor. When Ethel said, "I am the most alive dead one you ever saw," Thomas couldn't help rejoicing. In December, when the Rhines were struggling to recover hope about psychical research, John Thomas was bursting with it.

Thomas contacted the psychology department at Harvard, which put him in touch with Rhine. Thomas wrote that he was leaving for England to sit with a medium who didn't know him or anything about him. If the results he got with a medium there were as convincing, would J.B. and Louie be willing to work with him and help him? Without committing himself, Rhine nonetheless communicated his interest.

The results Thomas got from the British medium were again startling. It wasn't so much the individual revelations, like identifying an unusual necklace that had been buried with his wife. It was the sheer number of facts, which would go up into the thousands by the time he was done. "He came back a different man," Louie remembered. "He was the man who had lost his wife and now he found her again." He was no longer alone. "It had an effect on us too," Louie admitted. Maybe the results were genuine.

Through a British medium Ethel had urged her husband to see McDougall, who by this time had left Harvard for Duke. Thomas wrote McDougall that he had seven hundred fifty pages of material from all his visits with mediums and he wanted to do something with it. Thomas suggested that the Rhines come to Duke to work on his material in an advisory capacity, and then wrote J.B. to see what he'd do if McDougall approved. Rhine mulled this over and though he wrote back, "I am vacillating between a tremendous elation and a gloomy despair with regard to the whole field," at the end of the summer when Thomas wired Rhine with the news that McDougall had given his approval, the Rhines started packing their bags.

The very next month, September 1927, J.B. and Louie were installed in Durham. Once again, J. B. Rhine had pulled up stakes to answer what was coming to be known as the survival question: Is there life after death? Rhine was excited about Thomas's material, yet hesitant. The following spring, however, McDougall and Rhine had agreed telepathy was the most promising lead Thomas's results offered, and the place to pursue it was not in the ever-expanding cemeteries strewn across the country, but inside a laboratory.

TWO

A few months later, on the other side of the country, in Long Beach, California, Mary Craig, the frail wife of the writer Upton Sinclair, also began a series of experiments to prove the existence of telepathy. She'd had some psychic experiences early in her life and she was determined to find out if they were real or just her imagination. At a set hour, Mary Craig's brother-in-law Bob Irwin would make a drawing in his home in Pasadena and then concentrate on it for fifteen to twenty minutes. At the same time, Mary Craig would be lying on her couch in semidarkness trying to reach a state of what she described as an "uncomplicated" concentration that allowed her to receive a mental image of Bob's drawing. She would then make her own rendering to compare with her brother-in-law's. The experiment was refined in various ways; for instance, sometimes the drawings were placed in sealed envelopes that Mary Craig would hold against her body as she quietly waited for the visions to come "with a rustling of wings" and a feeling of triumph. She worked with several people, including Upton himself, and Roman Ostoja, a young medium whose convincing demonstrations of telepathy inspired Craig to begin her own experiments.

Mary Craig conducted two hundred ninety tests for nearly a year, and the results were impossible to ignore. A large percentage of the drawings she made were either almost exact replicas of the target drawings or inarguably close, and she could do it again and again. Sinclair, who described himself as "always interested, and always uncertain" about psychic research, was happy that fate had sent him a wife who practiced telepathy and who had the ambition to find an answer once and for all. He put together a manuscript describing his wife's investigations called *Mental Radio.* Part memoir, part science, the book ends with, "We present here a mass of real evidence, and we shall not be troubled by any amount of ridicule from the ignorant. I tell you—and because it is so important, I put it in capital letters: TELEPATHY HAPPENS!" Sinclair knew what he was in for. Anyone who lent credence to telepathy was subject to attack.

When word got out about Mary Craig's experiments a critical article soon appeared titled "Sinclair Goes Spooky," and this was written by someone whom Sinclair called a friend. Even McDougall, at fifty-eight, was starting to grow weary of the challenges to his standing due to his open interest in psychic research, and the repeated sting from men in more established fields. Just the year before, he had written in a letter to the *New York Times,* ". . . for thirty years I have taken an active part in the organized effort to obtain evidence on the question of life after death." This quest had "destroyed my reputation as a scientist" and exposed him "to the contempt of a multitude of the younger scientists of this country."

It was not a field to be entered into lightly. Scientists of the caliber and reputation of Thomas Edison, an American icon of progress and invention, had to confront disdain when looking at life after death. When his idea for a machine for communicating with the dead was published in *Scientific American* on October 30, 1920, the editors were so concerned for their own reputations they felt they had to explain themselves at the beginning of the piece in a sidebar that was outlined twice for emphasis.

Sinclair's agent wanted to protect his reputation, and he sought the support of respectable academics and scientists to give his manuscript some weight and credibility. Albert Einstein was a friend of Sinclair's and he agreed to provide the preface. Sinclair was basically fishing for blurbs when he wrote McDougall on May 8, 1929. The manuscript couldn't have arrived at a better time and into the hands of a more enthusiastic reader. Mary Craig's experiments encouraged him. He asked Sinclair if he could show the manuscript to Rhine, who "is just as keen on this sort of investigation as I am myself," and Sinclair wholeheartedly agreed.

McDougall wanted Mary Craig to repeat her experiments under the supervision of scientists. But some things cannot be performed on command, Sinclair tried to explain to McDougall. "The way these telepathy tests were done was always at a time when the mood happened to strike her, and they were always with persons whom she knew so intimately that there was no disturbance of her mood. To do the same thing with strangers would be entirely different." It was their first indication of just how tricky it would be to capture telepathy in a laboratory.

At the same time, a conflict was developing between John Thomas and Rhine, one that would be repeated throughout

Rhine's life. Thomas was convinced that the mediums' information was coming from his dead wife, Ethel. Rhine was still open to that possibility. Thomas's work was meticulous and exhaustive, and the mediums he had consulted supplied too much information to be explained away. But now Rhine had Mary Craig's experiments, which not only dramatically demonstrated that telepathy might be real, and therefore an equally possible explanation, they suggested how he might begin developing experiments of his own to prove it. Besides, he tried to explain to Thomas, even if the medium *was* getting her information from a living person, like Thomas himself, that wasn't such bad news. In order to confirm that we survive death, we have to find something of us that can exist independently of the body. If we can prove that telepathy exists, Rhine pointed out, that would be an important first step to answering the survival question. We have to find irrefutable evidence of telepathy before we do anything else.

For John Thomas, it must have felt as if Rhine had ripped Ethel right out of his arms. Proving telepathy would mean that all those messages hadn't come from his wife but were merely the medium reading his mind. "You can't lose me," Ethel had promised her husband via a medium. But Rhine was on fire. He had a real place to begin his life's work. Experiments would be designed to find out how telepathy worked. Until telepathy could be understood and ruled out, there'd be no way of knowing if the medium was getting her information from Ethel or from Thomas himself, and they would not have scientifically proven life after death. McDougall and others agreed with Rhine that this was the most scientific way to proceed.

But how to introduce this idea into a hostile scientific community? Rhine began to plan experiments in extrasensory perception, or ESP, a term he preferred to others used at the time.

"I am using the term 'extra-sensory perception' just now," he explained, "in order to make it sound as normal as may be." Rhine knew telepathy was going to be an enormous pill for science to swallow, and anything he could do to make it go down easier, beginning with a term that didn't have a hint of spiritualism or Ouija boards, served his purposes. Psychologists were accustomed to studying perception. Rhine hoped that using a term with the word *perception* in it would make it more palatable.

Rhine knew he had to design experiments that provided the least ambiguous results, and could be easily repeated in laboratories all over the world. The problem with Mary Craig's drawings was the subjective nature of analyzing the results. Her brother-in-law might draw a cactus and Mary would draw a flower that resembled a cactus the way she drew it. Should it be counted or not?

Rhine began with simple, relatively informal guessing contests among children at a summer camp. He'd hold a card with a number on it in his hand, and the children tried to guess the number. Their responses were either right or wrong. That fall he repeated the experiment with Duke psychology students, except now the cards were in sealed envelopes. They experimented with hypnotism to see if that enhanced the subjects' ESP, but the results weren't promising enough to continue. They did learn, however, that if they used a regular deck of playing cards, people showed a preference for one suit over another, or a particular card. There was no one department within the university that was absolutely ideal for parapsychological research, but the psychology department was where the research was welcomed, and when Rhine was confronted with the problem of people's associations with familiar playing cards, psychologists, it turned out, were uniquely suited for solving it. Rhine asked the psychologist Karl Zener for help, and Zener responded by designing a new deck of cards for Rhine's experiments. Each deck had twenty-five cards. The cards consisted of five sets of five cards with the following symbols: circle, square, cross, wavy lines, and star. People didn't seem to have a bias for the symbols Zener selected. There would come a time when Zener would be embarrassed about his contribution to Rhine's research and would insist that everyone stop calling them Zener cards. But for now Rhine and everyone in the psychology department were all cordial, work with the cards was progressing, and McDougall couldn't have been happier. "My colleague Rhine has been getting results that can only be called clairvoyant," he wrote Upton Sinclair that November.

A group of hardworking students were starting to gather around Rhine, including two quiet, earnest undergraduates: Gaither Pratt, a twenty-year-old student in Greek and philosophy, and Charles Stuart, a twenty-two-year-old who was majoring in mathematics. Charlie was a careful young man, gentle and scholarly, and fragile in both health and personality. Gaither Pratt was one of ten children, who had grown up on a farm in North Carolina. He was at Duke on scholarship and was washing dishes and typing his fellow students' assignments in order to meet his expenses. Like Rhine, Pratt had intended to become a minister but changed his mind when the answers religion held to the questions he was starting to ask weren't enough. Along with Rhine, the two young men started conducting ESP tests with everyone around them who was willing, and the tests were such a fun break from traditional studies that the Duke University students couldn't sign up fast enough.

In the beginning of 1931, while Rhine was refining his

experiments, an excited Sinclair wrote McDougall, "Einstein is here and we have seen a great deal of him—a very lovable person." Einstein was visiting Mount Wilson Observatory, where scientists had found evidence of cosmic background radiation, the first real proof that the universe was expanding. Einstein, in turn, wanted to show the Mount Wilson scientists the beginning of his unified field theory. The Sinclairs hoped to get Einstein to attend a séance with Roman Ostoja, the medium Mary Craig had worked with in her telepathy experiments.

Einstein accepted the Sinclairs' séance invitation and brought two physicists from Caltech, one of whom would go on to become a scientific adviser on the Manhattan Project. Gathering a group of scientists "who would make real tests" of Roman had long been a dream of the Sinclairs, and now arguably the greatest scientist in the world was about to sit down and witness what a real medium could do.

An informal séance wasn't exactly a controlled experiment and in any case, "it was a flat failure," Sinclair wrote Rhine in despair. "Roman pleaded that he had not done it for three years" and asked for another chance, but Einstein was too busy. McDougall, who had met Roman and described him as "very erratic," probably saw this as an important, and wasted, opportunity to make an impression on Einstein, who could have been an enormously influential ally.

In reality, there was never a genuine opportunity to persuade Einstein. According to Helen Dukas, Einstein's secretary, Einstein once said, "Even if I saw a ghost I wouldn't believe it." Helen claimed he went only out of friendship. "He was a sweet man." And it seems that what she said was true. Years later Einstein wrote to parapsychologist Jan Ehrenwald that he had supplied the preface for Sinclair's *Mental Radio* "because of my personal friendship with the author, and I did it without revealing my lack of conviction, but also without being dishonest." However, Einstein revised his opinion a few months later after reading Ehrenwald's book about telepathy. "Your book has been very stimulating for me, and it has somewhat 'softened' my originally quite negative attitude toward the whole of this complex of questions. One should not walk through the world wearing blinders." It's an indication that had Einstein witnessed a more sober, controlled, and persuasive demonstration of psychic abilities, parapsychology may have had if not Albert Einstein's support, his unwillingness to condemn and ridicule the investigations. That would have been as important as mountains of data.

Rhine would conduct ten thousand ESP trials with sixty-three students that same year, many of whom scored above chance and demonstrated ESP, but the following year Rhine found their greatest subject of all: a young divinity school student named Hubert Pearce. The son of a plumber, Hubert came from Clarendon, Arkansas—a tiny town that even today does not have a stoplight. He was as stable as Roman Ostoja was erratic, and Rhine and McDougall may have later looked back and regretted that Hubert hadn't been the one to sit down with Einstein that evening at the Sinclairs', instead of Roman.

Hubert had shown up at one of Rhine's lectures one day because he had a feeling he was telepathic, and the possibility scared him. He listened to Rhine's lecture about telepathy, then he waited in the back of the room for everyone else to leave. He was not only afraid to have his suspicions confirmed, he didn't want anyone else to know. Hubert was going to become a minister, and as his wife, Lucille, would later say, "Of all the places we ever lived, he never discussed it—a Methodist preacher believing in ESP—that was not a good idea." It wasn't that Hubert was ultraconservative himself, but he was only twenty-six and apparently ESP was even more dangerous than evolution. So Rhine grabbed some Zener cards that had been hastily hand-stamped by Gaither Pratt and went to Hubert's dorm room with a secretary to try a few tests where no one could see.

According to probability theory, just guessing randomly should turn up five correct answers for every run of twenty-five cards. If you get more than that, it's significant and evidence that something else is at work. Hubert Pearce was about to do significantly better than everyone else.

Rhine and Hubert began the first test. Rhine held each card facedown, under his hand. Hubert would make a guess about the card, and Rhine would turn it over. There was little conversation beyond Hubert's one-word guesses and Rhine's one-word responses. Rhine held down the first card and Hubert made his first guess. "Right." Hubert guessed again. "Right." Again. "Right." He'd ultimately score ten in a row. It was highly improbable that anyone would guess ten in a row just by chance, but improbable things happen. They just don't happen a lot. Could he do it again?

Rhine looked at Hubert. Intense concentration always defined Rhine's character, and nothing could break his attention once it was set. When Rhine was a twenty-three-year-old marine he won the President's Match, a high-power rifle contest in which all branches of the armed forces compete. A heavy thunderstorm broke out during the competition that year and drenched the marksmen. Rhine never even flinched. The rain poured down, the lightning exploded, and Rhine scored 289 out of a possible 300 and won. It cost him much of his hearing, but he won. And now that storm-defying, consequences-be-damned attention was focused on Hubert. Rhine's pale gray eyes bored into the young divinity student with something that must have been a pleasure for Hubert to see: expectation. Hubert was afraid, but who can resist the allure of such frank admiration and high-powered anticipation? Again, Hubert consistently guessed ten cards right for that deck, twice the number that would be expected by chance. Statistically speaking, Hubert was telepathic. Other students had demonstrated ESP, and Rhine already had a lot of evidence to make his case, but he knew that when going before the skeptical scientific community, the more unequivocal the subject the better, and Hubert's abilities were astounding.

Rhine also knew that a few informal tests in a dorm room weren't going to convince anyone. They had to repeat what they had done under strict and careful controls. Also, as impressive as Hubert's results were, all he and Rhine had done was find evidence of an effect. They knew nothing about it. What was ESP exactly, and how did it work? More experiments were needed.

Hubert started working regularly with Rhine and also Gaither Pratt, who was now a graduate student in the psychology department and working as Rhine's assistant. Eventually Hubert's fears were allayed enough for them to move from the dorm room to Rhine's office in the medical school building on Duke's west campus. For the next year and a half, Pratt and Rhine took turns testing Hubert. Unlike another promising student whose abilities had quickly faded, Hubert continued to score well for two years. Once he even produced twenty-five consecutive correct guesses. During one afternoon visit, Hubert wasn't performing well initially. As a means of encouragement, Rhine bet Hubert a hundred dollars that he couldn't guess the next card. Hubert got a hit. Rhine bet him another hundred dollars. Hubert got another hit. Rhine kept betting and Hubert kept getting them right until Pearce was owed twenty-five hundred dollars, and both men were trembling. Rhine never paid the twenty-five hundred dollars. Rhine and Hubert both said they knew on some level that the lab didn't really have that kind of money, but the fantasy worked.

Scientists would later express enormous skepticism at Rhine's ability to find good subjects. Very few following up on Rhine's results ever got anyone to score twenty-five out of twenty-five. But Gardner Murphy, a psychologist at Columbia University and a lifelong friend of Rhine's who was also conducting psychical research, had seen "the rugged force of the demands which he made upon his co-workers and subjects. In the light of his glowing intensity, it became possible to begin to understand the accounts given . . . of the way in which he had driven some of his subjects in the demand to get extrasensory phenomena." Rhine's colleagues always raised their eyebrows at the idea. Charisma? It was a quality they didn't know how to summon or measure and therefore preferred it was kept out of the laboratory.

But the betting experience with Hubert underscored something Rhine was beginning to understand about ESP, and that was when it comes to evincing ESP the subject's feelings mattered. When Rhine bet Hubert a hundred dollars it was during the Depression. Hubert was young and poor, and a hundred dollars was a fortune then. Even though deep down Hubert knew it was just a fantasy, human need can make much of fantasy. The times demanded the kind of hope and imagination that can produce results. A few years later Rhine would witness a similar demonstration of undisguised need. A nine-year-old from a local orphanage would score dramatically well in order to spend more time with a young woman at the lab she had grown attached to. "I am glad I got 23 right," the little girl wrote her. "But I want to get all of them right. I'm going to try very hard to get all of them right . . . I don't want to go home for I will miss you very much." When the child didn't do well the next time, she turned away from the experimenter with a serious and determined expression. "Don't say anything. I am going to try something." Wishing very hard the whole time, she proceeded to correctly guess all twenty-five cards. Unfortunately, unlike Hubert she couldn't keep it up.

Rhine recognized what was happening. But how do you control for something as variable and tenuous as feelings and need in a laboratory experiment? Rhine continued to refine their methods, and making their subjects at ease and as comfortable as possible in the lab and with the experimenters became part of their protocol.

Hubert continued to do so well Rhine knew they would be accused of cheating, so he began to add precautions. They put up a screen so Hubert couldn't see the cards. They used new decks for each experiment. In the fall of 1933, Rhine wanted them to see if Hubert's abilities were affected by distance. They decided to conduct tests while Gaither sat in a lab in the physics building on the west campus and Hubert sat in a cubicle among the stacks in the library, a hundred yards away. On the day of each test they'd meet in Gaither's room, set their watches, and then Gaither would watch from his window as Hubert crossed the quadrangle and entered the library.

At a predetermined time, Gaither would shuffle the cards and place the deck on a card table. Then he'd pick up the top card and place it down on the center of the table without looking at it. He'd wait one minute, and put it aside. He'd repeat this for every card in the deck. Hubert, meanwhile, was writing down his guesses, one a minute. When Gaither had gone through the entire deck, he'd turn all the cards over and note the order. They'd wait five minutes and go through the entire procedure again. When they were done they'd each seal their records in envelopes and separately take them straight to Rhine. For the first few tests, Hubert scored below his normal level, but after that he kept scoring higher and higher. They decided to increase the distance.

Hubert stayed in the library, but Pratt moved to Rhine's office in the medical school. They were now two hundred fifty yards apart. The results were mixed. Pratt went back to his room in the physics building and they tried again, but Hubert's scores were still mixed. There must be a psychological factor, they thought. If distance was a factor, all the runs and all scores would have been affected.

The results were still good enough that McDougall was afraid someone would accuse the young men of cheating. They did one last series with Rhine watching, to make sure no one left his station or tampered with the records in any way. In six runs, Hubert got high scores in all but one. By the time they were done with the long-distance experiments they had conducted 1,850 trials in all. They evaluated the results. The chances of Hubert getting the scores that he did for the series overall was one in an octillion. Rhine now had a subject he could introduce to the scientific world.

But then, just like that, it was over. Not long after the final distance experiment was completed, Hubert called to say he wouldn't be coming in that day. He'd gotten a Dear John letter from his girlfriend, and he wasn't in any shape for any experiments. Rhine and Pratt felt for him, but it didn't seem there was anything to be concerned about. Everyone's heart is broken at some point. They were about to learn that the heart was more crucial to success than they realized. Hubert's amazing abilities were lost. After that day, Hubert would never score as well again. He'd graduate the next year, go back to Arkansas, and almost lose his chance to become a minister owing to his work in psychology. He made it, he wrote them later, "but it did result in their sending me to a little country town up in the mountains, to a church that someone has let run down." In a few more years Hubert would send Rhine a picture of a pretty girl named Lucille. "I am enclosing a snapshot of the young lady who after January 19th will be my new boss," he announced happily. What meant the most to him in life had returned—love. Hubert would always be proud of his ESP accomplishments, and he kept in touch with Rhine and Pratt. For the rest of their lives the three men exchanged Christmas cards and visits and updates about their work and their children. Rhine and Pratt never gave up the hope that one day Pearce's abilities would return and neither did Pearce. If love came back, maybe his powers of telepathy would too. Years later, the three men would try again.

Rhine spent 1933 furiously writing up Hubert's astounding results and adding them to his book *Extra-Sensory Perception*, which would be published in the spring of the following year. But word about Rhine's work was already out, and he was becoming something of a figure around the Duke campus.

In April 1934 a well-known medium named Eileen Garrett arrived in Durham. She had offered to put herself completely at their disposal for any tests they wished, and everyone, especially McDougall, was thrilled. Ever since McDougall first read of Upton Sinclair's wife's experiments with Roman Ostoja, he had wanted to have a medium studied under laboratory conditions. Rhine couldn't wait to test her for ESP. Perhaps the same skills she'd developed as a medium would translate to even more astounding telepathic abilities than the ones previously displayed by Hubert Pearce.

Rhine had never met anyone like Eileen Garrett before. Whereas everything about Rhine was staid and contained, Garrett was unconventional and expansive. "She was a new type to all of us, an extremely vivacious Irish lady," Louie Rhine wrote later. "Even her red fingernails, commonplace now, were a sensation in our unsophisticated little community then." But Eileen was exotic to everyone. Aldous Huxley once said to her, "There are three creatures which really ought not to be: the giraffe, the duck-billed platypus and you, Eileen Garrett."

Rhine in particular couldn't figure out how to handle Eileen Garrett. Unlike everyone else around him, Eileen would never, not for a second, defer. Some claimed Rhine was attracted to Garrett, which may have been true. But it was also probably true that Eileen Garrett was simply too independent for Rhine. Although she married three times (one husband was killed in World War I and the other two marriages ended in divorce), Eileen once said, after shying away from yet another marriage proposal, that she finally realized "I would never make a good wife, but a pleasant mistress as long as I got my way."

According to her granddaughter, Garrett was one of the models for Patrick Dennis's famous character, Auntie Mame. Garrett had a publishing company called Creative Age Press, and Dennis did publicity and sold advertising for her. Apparently, she was as tough and demanding as Rhine. Hans Holzer had been dabbling with music and Broadway when he met Eileen, and she promptly told him to investigate ghosts instead. Holzer investigated ghosts for the rest of his life. "One could simply not say no to Eileen," he explained.

Eileen was forty-one when she came to Duke. Born and raised in Ireland by an aunt and uncle after her parents committed suicide shortly after her birth (Eileen's mother was shunned for marrying outside her faith), Garrett had an early life filled with all the good and bad elements of a fairy tale—misunderstood by a cold and unresponsive aunt and visits from people only she could see. Her perception was unique from an early age. She once wrote, "from the beginning, space has never been empty for me." Even when she was lying still in the sunlight, light was volatile—it moved and spiraled and burst, and she'd reach a state where everything had light, song, and sound. It's "almost electric in its reception," she wrote of the images that came to her. Upton Sinclair's wife, Mary Craig, also had to reach a certain state in order to see images, but while Mary Craig's condition was tranquil and sedate, Garrett's was a little wild.

Eileen went into her first trance in 1926, when she was thirty-three years old. While she was completely unaware, another personality who called himself Uvani took over and spoke to the people in the room. Uvani said he was there to manage the communication between the living and the dead, and to protect Eileen, who was vulnerable while in a trance. It was initially very frightening, and she pored through philosophy and psychology texts looking for an explanation of what was happening to her. She turned to other mediums and the British College of Psychic Science, which eventually helped her learn to accept and control her trances. Later, in 1930, another personality appeared who claimed to have died in Baghdad in 1229. For the rest her life, one or the other personality would continue to appear whenever Eileen went into a trance. Most of the people Eileen met in her efforts to understand the true nature of her trances accepted that the personalities were discarnate beings. But Eileen was not convinced, and by the time she came to Durham, she was as curious as Rhine to see what happened when he got her into the lab.

But Rhine and Garrett had a different take on everything to

do with psychic research. Their differences are probably never more clear than when you compare the publications the two would eventually separately issue, Rhine's *Journal of Parapsychology* and Garrett's magazine called *Tomorrow*. The *Journal of Parapsychology* published very dry, unvarnished research about ESP almost exclusively. Garrett's magazine started out as a literary magazine that also covered world affairs, but the first issue had articles like "Mediumship" and "Mysticism" and "Use Your ESP Every Day." A 1942 piece titled "Haunted New York" told the story of a young family who had moved into a Gramercy Park apartment, not knowing that a girl had thrown herself from the window of what was now their two-year-old son's bedroom. One night the child woke up screaming. When they ran into the room he was pointing to the window crying, "The lady jumped out."

Tomorrow was sensationalist, regularly covering topics like survival after death, hypnosis, fire walking, table tipping, the preservation of youth, crime, ghosts, fairies, witches, vampires, astrology, spiritualism, evil, falling stones, Lourdes, astrology, Atlantis, and just about every other paranormal topic that would never make it into the *Journal of Parapsychology*. There was a frequent focus on healing in the pages of *Tomorrow*, perhaps owing to Garrett's near-constant health problems, and while some of the articles seem a little silly today, a number of them, such as the ones on acupuncture, would now go under the heading of alternative medicine. From time to time *Tomorrow* also published articles about the psychic experiences of famous people, including Adolf Hitler, Mark Twain, Victor Hugo, Dwight Eisenhower, W. B. Yeats, Carl Jung, Thomas Mann, Sigmund Freud, Pope Pius XII, Abraham Lincoln, Percy Bysshe Shelley, Aldous Huxley, Albert Schweitzer, and William Blake.

But the mainstay of *Tomorrow* was its ghost stories. On the rare occasions that the *Journal of Parapsychology* included an article about ghosts they were called an "incorporeal personal agency" or a hallucination, and the pieces had titles like "Subjective Forms of Spontaneous Psi Experiences." (*Psi* is a catch-all term used for paranormal phenomena.) Articles in *Tomorrow* had titles like "The Ghost of Ash Manor," "A Grandmother's Unquiet Love," and "The Phantom Mistress of Rose Hall." Sometimes the titles read like lurid, paranormal *True Confessions* pieces: "Medium or Murderess?" "Search for Ecstasy," "I Don't Believe It, But It Happened," "How Did the Table Know?" "Apparition in Silk," "Séance Room Scoundrel," and "The Psychic and the Psychopath."

While Rhine and Pratt actually occasionally published in *Tomorrow*, some of Garrett's editorials read like a direct rebuke to their work. She'd acknowledge the importance of the lab's research and the statistical results they were getting, then publish an entire article devoted to arguing that it's time to move on from the Zener/ESP cards. The feeling was mutual. "It is probably too late now to answer your question about the Rockland County ghost story as it appeared in *Tomorrow* magazine," J.B. wrote Upton Sinclair. "The fact is, I have not read the story. I have been so disappointed in *Tomorrow* with its abandonment of standards and wide open reception to astrology and other uncritical forms of the occult that I have given the magazine little more than a glance."

But in 1934 there was enough mutual respect that the two were able to get past their differences, and in any case Rhine could not pass up the opportunity to study one of the greatest mediums of their time in a laboratory.

Garrett's initial impression of Rhine was similar to Gardner Murphy's, and she later said that her good results were directly attributable to the power of his enthusiasm. But she didn't like the ESP cards. Compared to seven-hundred-year-old dead men from the Middle East, the cards were lifeless and boring.

Rhine was constrained by the scientific method that he had to employ in order to have their work accepted. The problem was, "the ESP cards . . . did not stimulate my perceptions," Garrett complained. For her, the emotional component of her trances was essential to her success. But controlling for emotion in a lab was problematic, and for now the cards were all they had in the way of a simple, repeatable ESP experiment. Over seventy years later, the physicist Freeman Dyson would go even further about this dilemma. While freely admitting that he believes ESP exists, Dyson nonetheless concludes that emotion is so inextricably tied to ESP that a controlled scientific experiment for ESP is forever out of our reach. "The experiment necessarily excludes the human emotions that make ESP possible." However, Rhine himself was anything but lifeless and dull, and feelings don't entirely disappear when one enters a laboratory. Through the sheer force of Rhine's desire for Garrett to do well, after a couple of days, her performance improved. It would be an ongoing problem, but for now, by the time they were done, Eileen was scoring above chance in telepathy. Rhine also tested her for clairvoyance, but she performed less spectacularly. Telepathy (from the Greek *tele*, "distant," and *patheia*, "feeling") occurs when you get information directly from someone else's mind. Clairvoyance (from the French words *claire*, "clear," and *voyance*, "seeing") doesn't involve another person. You get information from somewhere else, such as holding an object that once belonged to a dead person. Using the ESP card experiments to illustrate, parapsychologist Nancy Zingrone explains it this way. "In Rhine's experiments, telepathy is when someone is holding the cards and another person guesses. Clairvoyance is when someone guesses while the cards remain in the deck."

Rhine later realized that the earlier card tests actually hadn't eliminated the possibility of clairvoyance and did not provide a true test of pure telepathy. The subject could have gotten the information from the cards and not the mind of the sender. In tests where they hadn't controlled for clairvoyance, Rhine would refer to the effect as GESP (general ESP) to distinguish it from true telepathy. For now it seemed that Eileen did better when getting information from people's minds rather than from the cards themselves.

Rhine also wanted to try experiments with Eileen's trances. Because her sittings were not only a scientific investigation but an entertaining break from the card tests, everyone who was anyone at the university wanted to participate and sit with Eileen, including the wife of the president of Duke University, along with Charlie Stuart, Gaither Pratt, J.B., Louie, Hubert Pearce, John Thomas, and some of Rhine's colleagues. A secretary recorded everything that was said.

Eileen would sit in a straight chair with her hands folded in her lap. She'd be quiet for a minute, then her head would go down in a sleeplike attitude. Her breathing would slow, sometimes she'd gasp, sometimes she'd utter a few low moans, then she'd raise her head and say, "It is I, Uvani." The voice and vocabulary would change, and she'd speak with a slight Eastern accent, but not very pronounced or theatrical. The sitters would then come in and take a seat behind Eileen.

Seventy-year-old transcripts of the Garrett sittings and the messages from Uvani read exactly like those of the well-known contemporary psychic John Edward and every other person who claims to talk to the dead today. It's as if Edward and others went back and read them for pointers. "I get the impression of two gentlemen and two ladies and I would like to tell you about the gentlemen first," Uvani would say through Eileen.

"The name William or Henry in connection with him." "He gives to me the sensation of holding up to me a chain . . . With this chain there is . . . It is like a locket, a pendant." Or "He was very tired and suffered from infection of the abdomen." Afterward Gaither made a list of all the facts that emerged from each sitting and asked all the participants to put a check next to the points they thought applied to them. There was a problem though. Rhine and Gaither realized that when going through the list, each sitter might have recognized things they heard in their sittings and put more checks next to those. In the next round the sitters sat in adjacent rooms, out of sight and out of the sound of the medium's voice. When they evaluated the results, the number of things Eileen got right were still greater than would be expected by chance. Unfortunately, as dramatic as Eileen's sittings were, the question about where she got her information remained. Was it from the dead, or through the living via telepathy?

Meanwhile, Rhine's colleagues in the psychology department had sent a letter to McDougall, who was in England. They were writing him because "grave dangers have arisen possibly threatening the integrity of our group." The danger was J. B. Rhine. His work had become so popular they feared it would soon completely overshadow their own. Without mentioning their communication, McDougall wrote an extremely delicately worded letter to Rhine, warning him to proceed more slowly and diplomatically.

In May of 1934 Rhine's book *Extra-Sensory Perception*, about his now ninety thousand ESP trials, was published. He'd already been attracting financial backing from outside the university, and one of his backers had arranged to have an additional four hundred copies sent to a list of Rhine's choosing. Rhine chose well, including eighty-five people from the most current volume of *Who's Who in America*, and his fame grew. Reviews started coming out that were not only positive, they were excited, and they were coming out in major newspapers all over the country, and then the world. Rhine was always accused of being a publicity hound, but in reality, he tended to view the media as a necessary evil. He wanted to get the word out that serious research of ESP was being conducted within a university, in order to attract other workers and to see who might contact them with stories and research of their own. But Rhine was also careful about what kind of publicity they received. When NBC called him about giving an ESP demonstration on the radio he turned them down. "I think you will agree with me," he wrote McDougall, "that this kind of thing is entirely out of the question." It would not be dignified.

Rhine soon found out about his colleagues' letter. Their actions made it all the more clear to him that he must separate himself to protect his research from the jealousy that could threaten their work. In earlier correspondence with the president of the university he had introduced the idea of setting up a separate Institute for Parapsychology, and it seemed the only thing standing in Rhine's way was money. The person who would end up making it all possible was Eileen Garrett, "the lady bountiful," as Rhine would later call her. As tough as she was, Garrett was something of a pushover when someone needed her help. Garrett introduced Rhine to her own and best financial resource, Frances Bolton, wife of the Ohio Congressman Chester C. Bolton.

Bolton wanted to help, but she made her particular interest clear—she wanted an answer to the survival question. Does something of us continue after death or not?

Rhine had to convince her that telepathy was the path to immortality, and he soon hit on the argument mostly like to persuade her. They couldn't study "poltergeists, various types of mediumship, and the like," from within the psychology department, he explained. If his colleagues couldn't handle a few decks of cards, what would their reaction be to the study of ghosts? "Before long there would be protests," he wrote. "People are afraid of the word 'spirits.' So we need the isolation to permit quiet work without feeling that we have to dodge attention." If he applied to a foundation, it would have to be on the basis of their ESP work and then he would be bound to work on only that. But if she could supply twenty-five thousand dollars, Duke would give him space for the lab, office materials, and a full-time secretary, among other things. "If you can make it possible, I will give it 10 to 15 years of my life—with all my energies."

Bolton agreed to give them ten thousand dollars a year for two years, and if all went well, she would continue to contribute that amount every year for up to five years. There would always be tension between Rhine and his backers. "They had little appreciation of the necessarily slow way that scientific investigations must proceed, beginning with the simpler elements of the problem," Louie later wrote. But in the beginning Rhine and Bolton were after the same thing, to show "that the human mind is not materially limited." At Bolton's request, her endowment would be named the William McDougall Research Fund, and her initial contribution would fund the early research of the lab.

Rhine was now forty years old, an associate professor, and everything was going his way. While Louie remained at home, conducting simple card tests with their children and their playmates, Rhine was out in the world making important connections everywhere, including Walter Kaempffert, a science writer for the *New York Times*, and others at *Scientific American*, where he was asked to be a contributing editor on the topic of psychical research. He was emerging as nothing less than an academic rock star. The funding he received that year from Bolton and others came to just under a tenth of the total research budget for all of Duke University. Whether he sought fame or not, it came after him with the full weight of an enormous public interest.

In the end, aside from a few resentful colleagues, the timing couldn't have been more propitious. It was 1935 and science had advanced beyond studying the observable world, to focus on the subatomic and invisible forces like electromagnetism. Everywhere, imaginations were ignited to what else might be out there unseen. The same year Upton Sinclair's book *Mental Radio* came out, Albert Einstein said, "The most beautiful thing we can experience is the mysterious. It is the source of all true art and science." So while scientists continued to be contemptuous of parapsychology, Rhine's experiments were having an impact, and many within the scientific community were willing to grant him at least some leeway to make his case. For the public, still thrilled by the invention of radio, whose operation was not understood and therefore something of a miracle, if you could send sound invisibly through the air, it didn't seem unthinkable that our minds could tune in to the voices of the dead, especially when this was exactly what they desperately wanted to be true.

People from all over the world began to write whenever they experienced something strange. Rhine even began a correspondence with the famous Swiss psychiatrist Carl Jung, who had received one of the four hundred copies of Rhine's book. Jung, who told Rhine that he was "highly interested in all questions concerning the peculiar character of the psyche with reference to time and space" and in particular, "the annihilation of these categories in certain mental activities," had always been open to psychic research. In his letter to Rhine, Jung described an incident from years before. He had begun holding séances with a young medium. Shortly after, a knife exploded inside a sideboard in Jung's house. Then a few days later a table spontaneously came apart. Jung believed the events were somehow connected to the medium.

Rhine was unused to having such easy acceptance of ESP from fellow psychologists, never mind someone of Carl Jung's stature. Rhine had to admit to Jung however that "we have no theory of mind [as of yet] which helps very much in dealing with these facts." Jung wrote back encouragingly. "There are things that are simply incomprehensible to the tough brains of our race and time. One simply risks to be held for crazy or insincere." Then Jung said what Rhine had been saying to Bolton about the need to lay low. "I have found that there are very few people who are interested in such things by healthy motives and fewer still who are able to think about such and similar matters, and so in the course of years I arrived at the conviction that the main difficulty doesn't consist in the question how to tell, but rather in how to tell it not."

But Rhine didn't think the problems he was having would be solved by keeping quiet about their work. The solution was gaining independence from those who would stand in their way. In September 1935 Rhine got almost everything he wanted. The university wasn't quite ready to give Rhine the full prestige of his own institute, but they were prepared to grant him independence. That fall Rhine was made the director of the newly christened Parapsychology Laboratory of Duke University. Everything was falling into place, and the future seemed so limitless it's not surprising that Rhine thought that after centuries of dreams, desires, and longings he would be the one to prove that maybe even death was not a real line of human demarcation. Specifically, Rhine and his colleagues were after definitive proof and labored to refine their experiments so that scientists all over the world could repeat their results. Parapsychology would no longer be a matter of believe/disbelieve, more faith than science. They were going to do nothing less than bring ghosts into the domain of scientific fact.

Rhine set out the lab's goals and defined four main areas of research: telepathy (getting information from other people's minds), clairvoyance (getting information from sources other than the mind, like objects), precognition (seeing into the future), and psychokinesis (moving objects with the mind). To keep the interest of people funding the lab, Rhine made it clear that he would also pursue curious incidents, but discreetly. "I would not dare be known as studying haunted houses," he wrote Bolton. But you can't keep things like haunted houses, ghosts, poltergeists, and demonic possession quiet.

THREE

Stacks of letters from around the country began to arrive at the lab daily. The mountain of publicity that followed the publication of *Extra-Sensory Perception* had established J. B. Rhine as America's authority for answers about anything paranormal. Among the flood of mail was an old-time-feeling but intriguing letter about a thirteen-year-old girl from Jonesville, Louisiana, named Alice Bell Kirby. Alice had special talents, the writer claimed, like levitation. "She just floats up like a balloon." She was also particularly adept at what the writer called "working the table." Without so much as a quickening of her heartbeat, Alice could raise a large dining room table with two men sitting on it who together weighed four hundred pounds. She just touched it with her fingertips, "and asks it to do what she wishes," which sometimes was to rise and sway with whatever music was playing. Once when visitors to Alice's house were leaving to go to town for a cold drink, the table followed them out the door and down the steps and into a car. Another time it climbed up the side of a wall. In addition to having these talents, Alice was also a trance medium, just like her grandmother (and Eileen Garrett).

Although everything about Alice's talents felt like a throwback to the last century's bag of psychical tricks, Rhine tried to keep an open mind. They were always on the lookout for authentic phenomena that might be tested in the lab, and this might be a genuine case of psychokinesis. While most people studying paranormal phenomena repeatedly went back to the same famous mediums, Rhine was testing college freshmen no one had ever heard of and getting demonstrably better results. In theory, there was no reason a child in Jonesville, Louisiana, couldn't display the same abilities as Hubert Pearce or anyone else he tested. Have her talents ever been witnessed by anyone with scientific training, Rhine wrote back. Does the table move without anyone touching it, in broad daylight?

The family invited Rhine to see for himself, but he was still skeptical enough to write Dr. Herman Walker of Louisiana State University to see if he'd be willing to make a preliminary investigation. Herman couldn't, but his wife Betty made arrangements to go with a friend.

Betty Walker and her friend Catherine Kendall went to the tiny town of Jonesville on a Saturday afternoon in November. Alice's teacher, who was also in attendance, said that Alice was a friendly, average student. Betty described her as a pretty girl, with wavy light brown hair, and gray-blue eyes. Betty also noted that Alice was "mature for her age. She is not very talkative but is very calm and over-sophisticated."

They gathered for the séance in a bedroom where the "performing table" had been placed. It was an ordinary room with a bed, dresser, wardrobe, and the table and six chairs. In addition to Betty and Catherine and Alice's teacher, they were joined by

Alice's older sister Rose, Mrs. Shelby White (the woman who had first written Rhine about Alice), and Mrs. White's son and daughter. A supervisor in the parish school system and his wife joined them after the demonstration started, along with a man to whom they were never introduced. Betty and Catherine seated themselves on either side of Alice, then everyone put their hands on the table, touching hands with the people on either side. "The lights were turned off and the room was in total darkness." Later, when the moon shot a ray of light through the window shade, they had to stop everything until quilts could be put up to block even that small sliver of illumination. "The table was asked to greet the newcomers but it refused. The lights were then turned on and the table was asked to explain its surliness." It didn't like the seating arrangements, it turned out. When everyone had been reseated, with Betty and Catherine no longer next to Alice, the performance resumed.

There were a number of special effects. The attendees asked questions and got raps on the table for answers, once for yes, twice for no. Sometimes the table would tilt and lift up about six inches. At other times both the table and chairs vibrated. Catherine felt a "girl's warm soft fingers" pass over her own hand three times. A little later both women felt a breeze. When a stool was placed on the table, Alice was lifted from the floor to the stool. Alice then told everyone that she had floated to the ceiling. Finally, with the lights back on, Alice wrote out answers to questions from everyone present. In order to demonstrate that the answers weren't coming from her she did this blindfolded. The whole performance lasted four hours.

Betty and Catherine wrote that by the time they got to the stool trick, "we were tired and more or less bored." They thought anyone could have manipulated the table. The man to

whom they were never introduced seemed to be "the most helpful of her friends." The breeze could have been caused by someone waving a handkerchief. Catherine's hand could have been
touched by Mrs. White's daughter, who was sitting to her left.
And when Alice announced that she had floated to the ceiling,
both Betty and Catherine immediately reached out and touched
Alice's shoes, which were still planted firmly on the stool.

Rhine accepted Betty and Catherine's version of the events
and his investigation ended, but several newspapers falsely reported that "Prof. J. R. Ryan [sic]" had been present and had
pronounced Alice "a marvel."

Immediately after, Alice's fame grew. People came in the
thousands at all hours from Mississippi, Arkansas, Texas, and
Tennessee. Offers to appear in nightclubs and onstage started
coming in, and people were always after her to make predictions about horse races and lotteries. But Alice refused, saying
she would not "commercialize her gift." The Universal Council for Psychic Research in New York telegraphed Alice's father
and offered ten thousand dollars if Alice could prove she had
supernatural powers, but the family did not respond.

Shortly after Betty's visit, the Columbia Broadcasting System
flew Alice, her mother, and a school superintendent to New
York City to appear on the radio show *We The People*. Alice answered the emcee's questions with an "unexcited calmness," a
local paper proudly related. She didn't know how she was able to
work the table, she told him. "I really don't enjoy this," she said.
"I wish I could go home and be a little girl again." When Alice
came home, a crowd of her neighbors, who had been there to see
her off, were there again at the airport to welcome her back.

Mrs. Shelby White kept trying for a while to interest Rhine
in studying Alice Bell, but his mind was made up. Young Alice
was nothing more than a Mina Crandon in the making.

But in that respect, Rhine was mistaken. Unlike Mina Crandon, Alice didn't spend the rest of her life touting her psychic wares. When she turned sixteen years old she decided she had had enough. "It was unsettling for a young girl. I felt like a freak," says Alice Bell, now an eighty-three-year-old widow with two sons and five grandchildren. It was always a spiritual thing for her, she explains. Her grandfather was a Baptist preacher, and while others thought it was the work of the devil, she chose to believe the opposite. Had she stayed with it, she believes, her abilities would have grown. But she wasn't comfortable with the trance part of her abilities. "It frightened me and I didn't like it." And she didn't enjoy the attention and publicity. The local kids were scared of her, and she wanted friends. Everywhere she went people would point and say "That's the witch." Alice put it behind her, grew up, and went to college to become a nurse. Soon after, she met and married an Air Force lieutenant and dropped out of school. The spirits weren't happy about her choice, and at the beginning of Alice's marriage they hit and shook the bed whenever the couple tried to sleep. But after a while the spirits gave up and started fading away. Alice was okay with that. "I always said, I just want to be normal, I just want to be normal. And after I got married I got normal."

Alice Bell now lives in Mississippi where she has a florist shop. On the phone she is endearing, lively, and funny, except for a couple of minutes when she talked about her husband's last day alive. He was frightened, and he asked Alice to appeal to her former spirit friends. "If they can help me," he pleaded, "now is the time." But Alice knew this wasn't something the spirits could do. She had to tell him that all her powers and afterlife connections couldn't save him. A few hours later he died.

Whatever the explanation for the strange events in her life, Alice Bell seems completely sincere. She wasn't faking it, she insists. People in her hometown still remember her amazing abilities. They look at her and ask, "Are you the one?" And she answers truthfully, "Yes." Two weeks after Betty and Catherine's visit, *Newsweek* sent a correspondent who telegrammed that he couldn't find evidence of trickery, and when the table rose and danced, "I firmly held her hands and despite pressure I exerted [the] table came up [and] pressed against my chest."

Rhine's goal was to apply the scientific method to the study of supernatural phenomena and have those results accepted by the scientific community. Any case he involved himself in had to be unimpeachable, and according to Betty Walker's report, Alice's power had been embellished. Rhine was already learning that it's one thing to find evidence for something; it's another to have that evidence accepted. As much as everyone outside academia and science seemed prepared to accept Rhine's discoveries as long as the experiments had been properly conducted, criticism from within the scientific community for his book *Extra-Sensory Perception* began almost immediately and grew. Psychologists and others went over his experiments in order to find faults and a reason to discount them, and unfortunately, Rhine gave them plenty to pounce on. He didn't describe the experiments in detail. He barely discussed the ESP cards that were essential to his entire thesis. The very thing that attracted students, his enthusiasm, annoyed his colleagues when it ran all through his book. He was not circumspect and humble, as a scientist proposing a new theory was expected to be.

For someone who didn't have a lot of friends or credibility in the field, he made almost every mistake he could have. To make matters worse, *Extra-Sensory Perception* was published not by a respectable academic press but by the Boston Society for

Psychical Research. Finally, his use of statistics provided so many possible points of contention that psychologists, mathematicians, and statisticians would argue about it for decades. But the academic combat only excited Rhine. "He has a temperament which needs a cause," Louie wrote about her husband. "He's a crusader." Rhine relished the battle.

In the beginning of 1936 Rhine made a triumphant East Coast lecture tour to eleven cities, and later that winter, Rhine's former professor, Dr. Edwin G. Boring, invited him to speak at Harvard. A two-hour presentation and discussion at Harvard led to a dinner that lasted until ten p.m., until Rhine had to get on a train to leave. "He made a big splash," one of the attendees, a Harvard psychology professor, told Gardner Murphy. "What does that mean?" Murphy asked, eager for more details. "I said just that, a BIG SPLASH. Everybody was tremendously impressed. They all want to know more about the subject."

As support for Rhine grew in elite academic and scientific circles, so did the amount of criticism. Professors railed against Rhine so unfairly, frequently attacking his work without really studying it, that students took Rhine's side and rallied to defend him. The often emotional response from Rhine's critics also brought Rhine an impressive group of professional champions, including Dr. Thornton Fry of Bell Telephone Laboratories. Sir Ronald Aylmer Fisher, one of the leading authorities in the world on statistics and probability at the time, wrote Rhine that he couldn't find any ground for the objections they'd been getting. "I should judge that your method somewhat exaggerated the significance of the combined result," he admitted, "but not to a sufficient amount to affect your conclusions." Rhine put the letter in an appendix in a later printing of *Extra-Sensory Perception*.

Help always showed up right when they needed it, it seemed. Charles Ozanne, an Ohio schoolteacher who had begun corresponding with Rhine, was now retired and at loose ends in his life. His mother had died, and he regularly visited mediums in order to communicate with her. Like John Thomas, Ozanne wanted Rhine to prove that the messages he was receiving were genuine. He sent Rhine the funds to establish an academic journal, solving the ongoing problem of where to publish after they concluded their experiments and wrote their papers. The first issue of the *Journal of Parapsychology* was published out of Duke University in April 1937.

That same year, Rhine began another lifelong friendship, with a wealthy, influential go-getter, Commander Eugene F. McDonald. McDonald was the founder and president of the Zenith Radio Corporation, and he wanted to do a series of ESP tests via the radio. Rhine had already turned down an earlier request for a radio demonstration, but Commander McDonald was harder to resist. He was big and husky, a man's man, like Rhine, and he genuinely liked and admired Rhine. McDonald flew the Rhines out to Chicago. On a Sunday evening in June, McDonald entertained J.B. and Louie on his yacht. Rhine and McDonald put their cards on the table. If he agreed to participate, Rhine wanted the experiment to be as scientific as possible. But McDonald felt that "we have a big job on our hands to entertain the American public by making them think." McDonald agreed that it shouldn't be some dog and pony show, however. It should be a real test, a nationwide test of mental telepathy via the radio. Rhine relented. He would participate but remain in the background, as a consultant. The tests would begin that September.

McDonald thought of everything. He put his attorneys to work on getting Rhine copyrights and a couple of trademarks for the designs on the back of his ESP cards (under Games, Toys & Sporting Goods), and negotiated a deal to start producing ESP cards to be distributed on a large scale. Then he arranged for Rhine to get royalties on every deck sold.

Rhine's life was a roller coaster going full tilt. His second book, *New Frontiers of the Mind*, from publishers Farrar & Rhinehart (a respectable firm), and the October Book-of-the-Month-Club choice, had been mailed out to friends and reviewers and was about to reach the bookstores. Right before the first Zenith broadcast took place, however, Rhine was attacked by one of his most vicious critics to date, Professor Chester E. Kellogg of McGill University. Kellogg wasn't satisfied with merely decrying the statistics, he accused Rhine of ruining lives with his work: "the public is being misled, the energies of young men and women in their most vital years of professional training are being diverted into a side-issue, and funds expended that might instead support research into problems of real importance to human welfare."

Kellogg's article was so overwrought that once again it brought Rhine as many champions as it did detractors. In this way the scientific community were consistent. They didn't like Rhine's exuberance for his subject, but neither did they approve of an overly vitriolic critic, and Rhine got a lot of letters of support. Edward V. Huntington, a professor of mathematics at Harvard and the first president of the Mathematical Association of America, wrote Rhine that not only was he not convinced by any of Kellogg's argument, "I do not like either its substance or its manner." Huntington drove down from Boston to Durham to study their research and assured them that their math was sound.

Rhine was discouraged by the Kellogg piece, but McDonald's right-hand man wrote to him to cheer up. "Nothing stops a crowd on a street like a fight." What mattered was that they had an audience.

The first Zenith broadcast went ahead on September 26, 1937. On a Sunday at ten p.m., in a Chicago studio, Zener cards were chosen one by one on a roulette wheel. Ten "senders" would concentrate on each card for ten seconds, before going on to the next. All of this took place behind a screen to ensure that no one could see what was on the cards, and the broadcast went out over what NBC called their Blue Network, which was used for news and culture programming (pure entertainment went out on their Red Network).

The novelty of an ESP broadcast turned out to be a welcome addition to NBC's regular Sunday night programming. Listeners were asked to mail in their guesses, and a week later Zenith had forty thousand responses to analyze. Woolworth's, which carried the ESP cards, sold out its entire stock after the first Zenith broadcast and had to place a rush order for more.

But the Zenith ESP cards had been printed on flimsy cardstock. Held in the right light and at the right angle, their symbols were easily readable. While the defect wasn't going to affect Rhine's work, because in his experiments either the "sender" and the "receiver" were not in the same room or there was a screen between them, B. F. Skinner, famous for his experiments that involved studying animal behavior in a box, now referred to as a Skinner Box, and a sharp critic of Rhine's, immediately discovered the defect in the cards. In a dramatic demonstration in front of his students, Skinner correctly "guessed" a hundred cards in a row. He wrote Rhine, in an affectedly casual manner, "I suppose you know about this, but we found it amusing. Do you plan to use data obtained with these cards?"

Rhine wrote back that he was aware of the defect. "You will

understand now why it is that we rely upon screening, sealed packs, distance, and the like, for our conclusions regarding ESP." But he admitted, "The defect in the printing of the cards to which you refer is one which has caused us, however, enormous misery." Zenith wasn't about to throw out the hundred and fifty thousand decks that had already been printed. Rhine then explained that they were working with the manufacturers to fix the defect. But this "is of no consequence unless, as I suspect is the case, you contemplate giving this news to the public."

When Skinner replied, all pretense at light amusement was gone. "I am rather surprised to find that you have known about this defect without yourself making it public," he wrote. "I did not intend to 'give the news to the public,' as you put it, although I did describe your work at a Student's Forum a week ago and read 23 out of 25 cards from the back, explaining how I did it." That was in addition to the demonstration of the defect that he had already given to his own students at the University of Michigan. "I have also written to Dael Wolfe at Chicago [a known strong critic of Rhine's], since he told me at the time of my review of your book that he was interested." Wolfe, of course, jumped on the news, and added Skinner's discovery to an article he was just then preparing about Rhine's work.

Skinner threw in one more insult, saying that this was "only another example of the kind of thing which is responsible for the failure of many of us to take your work seriously," before finishing with his most disingenuous statement of all: "I say that in all friendliness."

"Just how would you have had us inform the public?" Rhine wrote back. There was really no obvious vehicle for getting the word out. Moreover, these were not the cards that were supposed to be used by scientists for their tests. Rhine did, however, explain the defect in the next issue of the *Journal of Parapsychology*.

Over the next year, Rhine's positive reception in the popular press was matched by angry letters to the editor and critical articles in professional journals. Most of the critics focused on the statistics that backed up the lab's results, and their criticisms made it plain that many just didn't understand them. It's a misunderstanding that continues today.

Statistics are most useful when what you are looking at is rare, or the circumstances in which it arises are variable. That's why, for example, baseball is all about statistics. A player may not get a hit every game, but over the long run, a better batter will hit the ball more often than the next guy. ESP wasn't available on demand, but a very large number of trials over time would demonstrate its existence.

By 1940 the lab had conducted close to a million trials, and whether or not you call it telepathy, they had clearly demonstrated an unusual effect. If their experiments were properly designed, and the trials strictly controlled, their results can't be dismissed without throwing out the results of all experiments using the same statistical methods. That would include the methods pharmaceutical companies use to prove the safety of drugs millions of people take every day. The Duke scientists responded to every criticism of their controls and the design of their experiments, and successfully gathered sufficient evidence for the effect they called telepathy.

For psychologists who couldn't accept the results, the only way out was to continue insisting that "they must have made a mistake." Since these critics understood the statistics the least, that was usually their first target, but many of the attempts to replicate Rhine's experiments failed because they involved a hundred, or a thousand, or ten thousand times fewer trials.

By the end of 1937 the statisticians themselves had had enough.

In December 1937 Dr. Burton H. Camp and the Institute of Mathematical Statistics published a statement about the statistical side of the Parapsychology Lab's research. "Dr. Rhine's investigations have two aspects: experimental and statistical. On the experimental side mathematicians, of course, have nothing to say. On the statistical side, however, recent mathematical work has established the fact that, assuming that the experiments have been properly performed, the statistical analysis is essentially valid. If the Rhine investigation is to be fairly attacked, it must be on other than mathematical grounds."

Rhine was ecstatic. "The effect of this blast from the big guns of the mathematical world will, I believe, represent something of a turning point in our fortunes," Rhine wrote Ozanne. To Thornton Fry he wrote, "I feel that the issue is probably closed." Rhine couldn't have been more wrong.

Scientists around the country were repeating their experiments and some were getting similar results, but others who had not were using that to fuel their argument that ESP did not exist. What Rhine needed was a forum where he could confront the critics once and for all.

A few months later, Rhine received an invitation to participate in the American Psychological Association's (APA) symposium in Columbus, Ohio, that fall. The APA was *the* premier organization for psychology. Rhine wrote the president of Duke University that the invitation was "recognition indeed." A big conference, packed with psychologists from all over America—this was an opportunity to face their critics.

But a few days before they were scheduled to leave Rhine learned that evidence of cheating had been found in the results of someone who had successfully replicated their experiments, and someone planned to spring it on him at the conference. Rhine and his colleagues didn't know whose experiments were to be exposed, although they had their suspicions. However, there wasn't time before they left for Columbus to confirm them. When they arrived, instead of practicing their speeches they were holed up in their rooms going over pages and pages of original scoring sheets until they found that the results from a young student at Tarkio College in Missouri were just too uniform. The student's experiments were to be the centerpiece of Rhine's presentation. Now he had to scrap everything he prepared and wing it in front of an audience that wanted to see him fail.

The ESP symposium was held in a large auditorium in the chemistry building. The place was packed. The crowd was estimated to be at least five hundred, and people filled the aisles. Louie was terrified, but she made sure to sit still so that none of the watchful psychologists could see her shake.

A psychologist who had participated in the Zenith program went up first. Then Tom Greville, who was only twenty-eight years old at the time, was next. When he had agreed to be their math champion he assured Rhine that he would know everything critical that had been written "backwards and forwards," and everyone who had attended from the lab would later agree that he was the most poised and self-assured. Then a critic got up and once again pointed out the defect in the commercial ESP cards. Rhine was next. It was very hot that day. Rhine was wearing his heavy suit, and the windows did not open. There was no water on the podium, and by the time he asked for a glass and someone ran off to get it, he couldn't wait for the young man to return. He was forced to begin. Rhine was extremely nervous and agitated, and for the first time, he had trouble speaking. "I don't know why I should be so nervous," he said to the audience.

It was the worst speech he ever made, Louie would later say. But he gave up pretending to be cool and calm; he was a wreck and he surrendered to it. He was rattled, but he hung in there. His courage was plain, and so was his integrity. Everyone could also see that he was telling the truth. He may have had to struggle to do it, but he addressed every point, simply and unemotionally and without embellishment. People couldn't help rooting for him. He had a reasonable, nondefensive answer for every criticism, particularly the defect in the cards.

The members of the APA still didn't know what to make of the ESP experiments, but they could see that J. B. Rhine was neither a liar nor a fraud, and that much of the criticism they had just heard had been unfair. When Rhine finished he got more applause than any other speaker at the conference. Many years later, the well-known science writer and skeptic Martin Gardner would write a book called *Fad and Fallacies*, and say of Rhine, "It should be stated immediately that Rhine is clearly not a pseudoscientist to a degree even remotely comparable to that of most of the men discussed in this book. He is an intensely sincere man, whose work has been undertaken with a care and competence that cannot be dismissed easily, and which deserves a far more serious treatment."

In the end, no one brought up the falsified reports. After all the speakers had finished, the floor was opened up for discussion. The most satisfying response came from a former skeptic, Professor Bernard Riess. Someone had complained that Riess had rushed the results of his ESP experiments into print, and Riess directed his comments to him. "I undertook the experiment as a way of demonstrating to my classes that ESP did *not* occur," he countered. "I did not succeed in that. I do not know whether Dr. Britt believes in throwing away good data just because he doesn't precisely understand the full implications of that data, but I felt they should be reported." The audience applauded.

Afterward Rhine wrote everyone of their triumph, describing again and again the applause given to Riess and others. To one person he wrote that "the critics put up a poor case indeed." And to another, "At no time did the opposition get a foothold or score a point. Everything was answered promptly and with a bang."

Rhine's arguments were so sound the critics had nothing left to say, but this didn't mean they accepted the lab's results. Worse, while the Duke scientists had made some leeway with psychologists, they had made virtually none in other areas of science, and most particularly among physicists. When Rhine said ESP was evidence of something that operated independently of the body and was not in any way physical, he essentially cut off any possibility of general acceptance from physicists, who by definition are solely concerned with the physical world.

It was especially maddening when a critic would attack a point, like the defect in the cards, either not knowing or not caring that someone else had pointed it out, and that the lab had already responded. It would keep an issue alive that had already been addressed and force the scientists to fight the same battles again and again. The magician James Randi, who has made a career out of debunking the paranormal, says believers are like "unsinkable rubber ducks." But the same observation could be made about those inclined to reject the parapsychologist's findings. No matter how many modifications were made to address their complaints and improve the Duke experiments, they would never be accepted by some equally unsinkable skeptics. For them, Rhine and his ESP experiments were just as impossible to take seriously as Alice Bell Kirby and her dancing table. But Rhine and his staff didn't know this in 1938. After the conference in Columbus they were confident that a parapsychological revolution was just around the corner.

Now that the lab had gathered evidence of ESP, the next step was developing a theory. A working theory is even more necessary when you're talking about something like ESP—"the presence of theories facilitates our acceptance of implausible facts," the late Marcello Truzzi explained in *The Reception of Unconventional Science*. So the staff at the lab tried every test they could to learn more about ESP, and analyzed the results from different groups including children, the blind, American Indians, patients in mental hospitals, and animals. Animals were useful because they couldn't be accused of trying to defraud the experimenters. The results of the animal studies varied but nevertheless provided evidence of ESP. Still, whenever someone wants to portray J. B. Rhine in a less than flattering light, they bring up Lady Wonder, a telepathic horse that Rhine had once pronounced authentic. Some alleged later that Lady's owner had been giving her signals, but Rhine had taken precautions against fraud that included testing her without having the owner present, and Lady was still able to perform.

While the lab continued to pursue the elusive theory of ESP, something that would explain these abilities that appeared to operate independently of the body, one of Rhine's first champions was not doing so well. McDougall had stomach cancer, the second leading cause of cancer-related death in the world. When Upton Sinclair sent copies of his latest book to Rhine and McDougall, Rhine was forced to write back, "I am afraid Dr. McDougall will not get to read this. He is taking morphine every four hours now and knows that his time is short." By the end of November, the morphine was losing its potency. Rhine wrote that it was a blessing when McDougall slipped into a coma days before his death on Monday, November 28. An important protector and friend was gone.

While things were otherwise generally going well at the lab, there were indications that their bright world was darkening. At some point every day, Rhine turned his attention to his enormous correspondence, which included hints of the growing problems in Europe. A couple of weeks after the Columbus symposium, a letter arrived from a twenty-six-year-old psychologist in Berlin named Lilli Guggenheim. She had read Rhine's second book, *New Frontiers of the Mind*, and she had an idea. She suggested a way of using the Rorschach Ink Blot test to find good subjects for telepathy. However, as a Jew in Germany, she explained, she was excluded from scientific laboratories, so she couldn't do the tests herself. But if he gave it a try, "I am looking forward to hear some news about the results." Later, she wrote again and asked about a job at Duke. She was anxious to get out of Germany. Rhine wrote Don Adams, a professor from Duke's psychology department, to see if there was anything they could do for her, but they couldn't come up with a position. Lilli Guggenheim would be deported from Berlin in November 1942 and sent to Auschwitz, where she died.

World War II was coming and, by sad necessity, so was a reawakening of the public's interest in life after death.

FOUR

W e don't really know what we are." Written in 1944 as part of a research proposal, the line was meant as an intellectual call to arms, but given the time period and the lab's progress, it also has the feel of an unconscious confession. Scientists at the lab had been getting promising results from a number of experiments in psychokinesis, but they held out from publishing. If critics were continuing to have trouble accepting telepathy, how were they going to react to the idea that people could move objects with their minds? America had been at war for three years now, funding for the lab was at an all-time low, and uncertainty permeated not just the question of death, but every aspect of life. "These are dark days," Rhine wrote one of the lab's funders, "not only for research such as ours, but to civilization itself." There had been so much death that by June, sales of Ouija boards, which had been almost zero in 1943, went up to fifty thousand in one New York department store alone. In the small world of the lab, John Thomas, the man who had brought the Rhines to Duke, had died a few years before. This was after receiving a message from his dead wife, Ethel, "My position to him is altered. I am nearer to him. I am nearer to him than I have ever been before." Duke University president William Preston Few, their last remaining champion, also died that year. With both Few and McDougall gone their position at Duke was now precarious. A new president who wasn't open to parapsychology might decide to shut the lab down.

In so many other ways, the forties had started out well. In their first five years the staff at the lab had tripled, and by the beginning of 1940 they had fifteen full- and part-time workers. Every week they tested up to a hundred people for ESP. Charles Ozanne had just made the lab the beneficiary in his will. And Rhine's fourth book, *Extra-sensory Perception After 60 Years*, a team effort written by the lab staff for the scientific community, had come out in the spring, following a nine-page spread about their work in *Life* magazine. In it they tried to make up for all the mistakes Rhine made with his first book, and there were indications that they had succeeded. Rhine's old professor at Harvard, Edward Boring, wrote to tell him that parts of it were required reading for the Introduction to Psychology class, which meant that every young psychology student at Harvard would be familiar with the work of the Parapsychology Laboratory.

But three days after the Japanese bombed Pearl Harbor, Gaither Pratt wrote a colleague that he was eager "to do something a little more immediately related to the present world crisis than ESP," and put in an application to the Navy. Rhine, a former marine, tried to enlist as well, but he was turned down for physical reasons. Charlie Stuart knew no one was going to take him. He had a heart condition that made him ineligible. But when America went to war, people curtailed their investments in parapsychology, and so Charlie took a position at Stanford at the height of the war years. With the men away, the lab began to fill with women. Among them were Louisa Rhine, who was now working there part-time answering the daily mountain of correspondence; Dorothy Pope, who started out at the lab as a ninety-dollar-a-month secretary and eventually became the editor of the *Journal of Parapsychology*; and "the Bettys," Betty Humphrey, a philosophy major from Indiana, and Betty McMahan, a small-town girl who had transferred from Appalachian State Teachers College, and who was nicknamed BettyMac to distinguish her from Betty Humphrey.

The presence of the women in the lab changed everything. The affection that had always been there, but largely repressed, blossomed. Gaither Pratt and Charlie Stuart, for instance, had always addressed their letters very formally to "Dr. Rhine." Betty Humphrey, however, addressed hers with appellations like, "Dear Puny, I mean Bully," or "You Poor Little Folks." The 1940s were when the small group at the lab truly became a family.

There wasn't a lot of money for new experiments, so Rhine threw himself into evaluating the results of the experiments with psychokinesis (PK) that they'd been conducting since 1935. "If the case for PK is as good as I think it is," Rhine wrote, "we have here an explosive idea that will reach further than any of Hitler's bombs, but I know only too well that it will never explode in the right time or place unless Hitler's are stopped soon." The war was growing worse and Rhine knew he was never going to see combat. In 1943 the lab published eight papers and four editorials representing nine years of work in psychokinesis. Rhine called PK the "twin sister of ESP." It was the next logical step in proving the mind's independence from the body and therefore life after death. If the mind was powerless to affect the material world after the brain died, life after death was not possible. "But if the psyche is a force—in its own right, with laws and ways peculiarly non-physical, the survival hypothesis has at least a logical chance. If the mind is different from the physical brain system it *could* have a different destiny."

The PK experiment that the lab ultimately devised was inspired by a gambler who had visited the lab and said he could affect the fall of the dice with his will. "The visitor went on his way, but he left this beautifully simple idea behind," Gaither wrote. It was perfect: Roll a pair of dice and one either did or did not get the roll that was tried for. The results were simple and unambiguous.

Using regular six-sided cubes, they began testing, first themselves, then Duke students, and finally anyone else who was willing. In the beginning the subjects rolled the dice from cupped hands, then they bounced them off the wall, and then they rolled them down a chute. Eventually they built a machine to mechanically throw the dice. J.B. and Louie announced in their first report that "the experiments on PK show first that the mind has force, real kinetic force, and that it can also operate outside of the body."

But whenever someone brought him a dramatic example of that force outside the lab—a poltergeist—Rhine would take his usual initial skeptical stance. They couldn't get poltergeists into the laboratory to study them, and while Rhine believed that the "human mind can escape physical boundaries under certain conditions," he couldn't prove it with a ghost. Besides, the lab was still struggling to be taken seriously, and Rhine knew the scientific community would be more open to a simple experiment with dice than a case study with a poltergeist. But like telepathy, so far the experiments had established only that there was a PK effect. They still didn't know much about it. What kind of force was it, and how did it work, exactly?

It also wasn't long before they had the same problem with the dice that they had with the cards—people got bored. The experimenters had to make it more engaging. A limited budget and a country full of women and children due to the war inspired them to make the tests a game suitable for a children's party. They designed a series of experiments that came to be known as the PK Parties. The Rhines conveniently had a pack of young daughters—Sally, Betsy, and Rosie—who had a bunch of friends.

BettyMac designed and built an apparatus to shuffle small, nickel-size plastic disks that she'd painted red on one side, blue on the other. All the girls took turns at the machine. They'd concentrate hard on one color or the other, trying to get the machine to send out more disks of that color. "Think red, think red," the girls would whisper. If they got above chance results, they all got candy and gum and a small prize. This was more fun than ESP cards, and the results were encouraging.

Work moved slowly for the next couple of years as Rhine and his staff tried to learn what they could under the reduced circumstances imposed by a raging world war. They found that alcohol and caffeine had the same effect on PK abilities as they did with ESP. People did worse when a little drunk, and better with a little bit of a jolt. Testing revealed that PK, while there, was weak. No one was going to beat the house in Reno on the basis of their PK abilities.

Many years later Helmut Schmidt, a German physicist who worked at Boeing's research laboratory, and later for Rhine, would develop a PK experiment involving a device called a random number generator. Using the process of radioactive decay to randomly illuminate a circle of lights, subjects were asked to influence which direction the lights lit. The results were so impressive that when referring to later versions of this test, the scientist and skeptic Carl Sagan would grudgingly concede that "by thought alone humans can (barely) affect random number generators in computers," unable to resist pointing out that modern PK tests, like the telepathy tests, indicated a weak effect, as if weak effects are trivial or unimportant. The effect of aspirin in reducing heart attacks is even weaker but nonetheless life-saving. The fact that the effect of PK is weak does not in itself diminish its significance.

The war continued, and letters arrived from inconsolable and misguided widows who hoped that Rhine had something better than a Ouija board for bringing their dead husbands back, if only to exchange a few more words. It must have been difficult for Rhine, the ex-marine, to disappoint them. The war brought other poignant correspondents, like Harold Scharper, a soldier who was wounded and returned home a paraplegic. He read one of Rhine's books and was inspired to offer himself up to Rhine for research, suggesting that disabled men could be a valuable resource as subjects in paranormal testing. ESP and PK would be "an asset to most disabled people for we have the time and can develop the patience needed far better than the normal person." Harold firmly believed that "the work that you are doing can help me and others like myself to become leaders in many fields."

Rhine, who believed more in the abilities of the mind than in the afterlife, wrote a tender letter back. He assured Harold that it was true "about the capacity of the mind to make up in one for the handicaps that may develop," and that "our research here does, I think, help support the view that the mind is a free, creative, volitional system. It does encourage one to reach out for greater powers. It adds to the sense of adventure in merely being alive."

Although it looked like the war would never end, when science and government are equally determined to make something happen, it happens. The Manhattan Project was started in 1941, and four years later, on August 6 and 9, 1945, atomic bombs were dropped on Hiroshima and Nagasaki. On August 15, Japan announced their surrender to the world.

Slowly the men returned and the lab put itself back together. The full-time staff at the lab was now Gaither Pratt, Charlie Stuart, the Bettys, and six part-time graduate students. They were having a space problem for the first time since 1942, and it would stay this way for several years until Charlie, always frail, succumbed to sickness early in the winter of 1947. His passing brought the staff still closer.

In the second half of the decade, the media once again turned its attention to the lab and to the paranormal in general. The *New York Times* wrote up the lab's experiments with psychokinesis, and Gian-Carlo Menotti's full-length opera, *The Medium*, premiered in New York at Columbia University. It told the story of a deaf mute medium who started out as a fraud but then developed genuine abilities that destroyed her.

At the end of 1946, Rhine wrote Dr. Margaret Mead at the American Museum of Natural History in New York. He wanted to know if she had come across any parapsychological occurrences in her anthropological studies, and if she had, would she be willing to write about them for their *Journal of Parapsychology*. Mead said that she was sympathetic to psychical research, but "from the sort of accumulated hearsay evidence which one collects, my feeling is that individuals with cultivated extra sensory capacities are about as rare in these primitive societies as they are in our own." She had been keeping her eyes open for them, though. "In Bali we were alert for possible manifestations but found none. Among the Iatmul of the Sepik river, claims to special knowledge are made for prophetic dreams but as these are never told until after the event which they are claimed to have prophesied, there is again, no data."

The next year the lab was bustling with activity when the Duke administration, including the new Duke president Robert Flowers, unanimously adopted a motion to remove the lab from the psychology department entirely and set it up "as a separate and independent agency to be known as the Parapsychology Laboratory of Duke University with Dr. J. B. Rhine as its Director." It was an inevitable change, and in some ways it was a change for the best. Martin Gardner would later write, "There is obviously an enormous, irrational prejudice on the part of most American psychologists . . . against even the possibility of extra-sensory mental powers." Encountering that prejudice was distracting, and the occasional confrontations sapped the researchers' strength. Rhine must have also seen the separation as raising their status in the university. They were no longer a subdepartment within another department, but a wholly independent lab. It wasn't the institute he dreamed of, but it was a start. The lab paid a price for its independence, however. It may have strengthened the staff's bond to each other, but it further isolated the lab from the rest of the academic community at Duke.

Once again the world was full of possibility. Sources of funding for research in nonstandard areas of science were beginning to open up. Alfred Kinsey had just founded the Institute for Research in Sex, Gender and Reproduction at Indiana University, and Rhine was corresponding with Dr. Alan Gregg, the man at the Rockefeller Foundation who had approved the funding for Kinsey. At the same time, however, Charles Ozanne, who was now regularly contributing to the lab, was expressing increasing dissatisfaction that they were not more directly attacking the question of life after death. Yes, their experiments were intriguing, but Ozanne was desperate for traction on the original question: Is there anything of us that survives death?

Rhine soon made an announcement that would appease him and his other funders. Rhine decided that the lab staff were going to start looking through the thousands of letters in their files in order to do a formal study of what they called "spontaneous psychic experiences." This was the term the lab used for "ghost stories." The announcement didn't mean that Rhine had changed his position on ghosts and poltergeists. If it couldn't be brought into the lab, it didn't exist for him. But Rhine believed the letters might provide clues for experiments. Perhaps there were spontaneous cases that could lead to that single answer that Ozanne and others wanted so badly. Taking personal experiences as a starting point, the lab staff could devise methods of verifying psychic events and studying them. But going through their enormous collection of letters would be a huge undertaking. At one of their regular Monday night meetings Rhine asked for volunteers. No one raised their hand.

When Louie went back to the lab part-time during the war years, she wrote her sister that "fifteen odd years of motherhood and domesticity leave me feeling considerably unscrewed." She was ready be a scientist again. But to Louie that meant doing experiments. She had worked with children here and there, and she wanted to resume research. Like Rhine, she didn't believe in ghosts, so the spontaneous cases were hardly the return to hard scientific work she dreamed of. But Rhine wanted it done. And so Louie put her own interests aside for her husband, and the Parapsychology Laboratory announced their new project to the world: The Spontaneous Case Collection, headed by Dr. Louisa E. Rhine. The laboratory was officially investigating ghost stories.

FIVE

A small bottle of water rises from a table. For a moment, it's enchanting. An ordinary object is taken from the mundane to the magical. Before anything else happens, before any opinions have formed or disbelief sets in, the world is more fantastic than it was seconds before, and everyone is mesmerized. Then it's launched at someone's head. Reality has now not only betrayed the witnesses, it has attacked them.

"Okay, did that just happen?" anyone might ask, "or am I insane?" When unseen forces start throwing things around, neither possibility, nor any other explanation for that matter, is comforting. While most people welcome the idea that death is not the end, and that we are not alone, the otherworldly beings they dream of, are raised on, and pray to in times of crisis are not malevolent spirits or demons hell-bent on destruction. The best anyone can hope for when poltergeist-like activity takes over a home or a person is that someone is playing a trick. Personal possession is a last and frightening explanation. The solution for possession, an exorcism, can be deadly.

In 1980 a woman poured scalding water over her twenty-month-old baby and then put him in a preheated oven. She was trying to rid him of the devil. Three years earlier two priests and a West German woman's parents were convicted of manslaughter when the young woman died of starvation on the very day that her demons were finally banished after more than six months of weekly exorcisms. An eight-year-old autistic boy in Wisconsin was asphyxiated during a botched exorcism performed by faith healers in 2003. The minister appealed his felony child abuse conviction, but it was upheld. In 2005, a Romanian monk and four nuns were charged with the illegal confinement and murder of a nun who was starved and crucified. "God has performed a miracle for her," the unrepentant priest proclaimed. "Finally Irina is delivered from evil." All five were convicted of manslaughter two years later. The priest appealed the verdict and lost, and news reports indicated that the nuns would appeal as well.

One day, while going through their daily correspondence with an eye for possible cases for her Spontaneous Case Collection study, Louie opened a letter from an anxious priest in Washington, D.C. The Reverend Luther Schulze wasn't thinking possession when he wrote Rhine for help. He believed a poltergeist was the more likely explanation for the disturbances centering around Roland (not his real name), the thirteen-year-old son of a Maryland family in the church's congregation. There is some debate among parapsychologists about just what a poltergeist is, a spirit or someone with PK abilities strong enough to move considerably more than a pair of dice, but in either case, for the most part poltergeists are benign. They may break things, but they aren't out to truly hurt anyone.

The same cannot be said about demonic possession. Once the devil or one of his disciples has a body it has essentially declared war on that body's well-being. By the time Schulze wrote Rhine, the Maryland family had given up on the idea that a poltergeist was responsible for their problems and had taken Roland to the Jesuits in St. Louis for an exorcism. The one thing most poltergeist cases have in common is that they are short-lived. It's usually only a matter of months before the disturbances stop and everyone separates into the real, imagined, or faked camps. Of course the Maryland family didn't know that it would shortly end. For all they knew this was how the world would be now, violent, scary, and apparently determined to hurt them. The newspapers would get hold of the story a few months later, and a college student named William Peter Blatty would read their accounts and years later turn them into a best-selling book and the blockbuster movie *The Exorcist*. Blatty would change the age and gender of the child and other details, but much of what Schulze related in his letters to Rhine is familiar.

"It first appeared on January 15, 1949," Schulze wrote Rhine in March. He then described what was later depicted on movie screens across America: shaking beds, objects flying through the air, heavy furniture sliding across the floor, words appearing on the boy's body, the ability to speak strange languages, and visions of the devil. The family was staying with relatives in Missouri, and on the day Schulze wrote Rhine their son was being attended to by priests in the old psychiatric wing of the Alexian Brothers Hospital in St. Louis. Schulze was trying to persuade the family to bring the boy home and to the care of a sympathetic physician. He reached out to Rhine for support. "Would you or someone from your staff be interested in studying this case?" he asked Rhine. Perhaps Schulze, a Lutheran, did not have much faith in the Catholic Church's approach to the problem.

Rhine could sometimes be as biased and unyielding about phenomena like poltergeists as scientists were toward him and ESP, but this was one of the most dramatic stories anyone had ever heard. Rhine was out of town, and so it was Louie who immediately replied, speaking for her husband. "I am certain that he will be intensely interested in the case you mention. I am sure that he will want to study the case, if possible, at the earliest opportunity." Rhine wrote himself a couple of weeks later to say that what Louie had written was true. He was interested. While he wrote that it was most likely that the boy created the effects, for Rhine, a dramatic demonstration of psychokinesis was better than a demon, and he was clearly excited. He asked Schulze to telephone him and reverse the charges if the boy returned and the phenomena recurred.

The family was already back home when Schulze got Rhine's response. Jesuit priests had been trying for weeks to exorcise whatever had taken possession of the boy but were so far unsuccessful. Although Rhine had agreed to investigate, Schulze told him that in spite of the poor results, the family was still leaning toward a religious explanation. They were very skeptical about anyone from a university. While they were in St. Louis a member of the Washington University faculty had told them "that some day we could tune in any of the departed we wished just as we now tune in a television program." This was not, apparently, what they wanted to hear. According to Schulze a mental health professional who had examined Roland had said "the boy did not seem to want to grow up and assume responsibility, but preferred holding on to his childhood." They didn't like the sound of that either. For the Maryland family, demons made more sense than science. Or maybe it was that religion offered more hope for relief. There would be no meeting with Dr. Rhine.

Instead, they took Roland back to St. Louis and the fifth floor of the Alexian Brothers Hospital. The exorcism would resume. Schulze was still skeptical. "I have a feeling that this family will require treatment as individuals and as a family eventually and that we will be expected to pick up the pieces." Rhine was still concerned that trickery, whether conscious or unconscious, might be the cause of the events, and he asked Schulze to describe what he himself had witnessed.

Schulze told Rhine about events on the evening of February 17, when Roland spent the night in his home. They both had gone to bed and were sleeping in twin beds in the same room. Should anything happen, Schulze would be there to see it. Around midnight, Roland moved from the bed to a chair because the bed at Schulze's house was trembling just like the one at home. Schulze was awake now too. After a short time the chair, with Roland in it, started to glide backward toward the wall. Schulze went over to the boy. Roland took his feet off the floor and placed them on the edge of the chair. Schulze watched as the boy and the chair slid all the way back against the wall and toppled over with the languid slowness of a dream. Rhine wondered why Schulze stood there and made no attempt to catch the boy, and Schulze tried to explain his paralysis. "I wanted to see if it would really go over." It was just so absurdly impossible he was transfixed. The truth was, curiosity had gotten the better of him and he had to see if the supernatural performance would continue. This was a big, heavy chair, Schulze added. "With a low center of gravity." When Schulze sat in it himself and rocked with all his might he couldn't get it to tip over. None of the furniture was going to sit still, it seemed, so Schulze made a bed of blankets for

Roland on the floor. Again, while he looked on, Roland and the pile of blankets slid from the center of the room to underneath the bed. It did this twice. The boy's body was rigid, Schulze told Rhine, and there was "no wrinkling of the bedding" to indicate that he was making any kind of motion that would explain the sliding. Since psychiatry couldn't convincingly explain that kind of movement, and he wasn't inclined toward belief in possession, the poltergeist explanation was all Schulze had left.

It's a remarkable series of letters. Schulze is not trying to persuade Rhine of anything, nor does he entertain fixed beliefs about the cause one way or another. He's simply searching for an explanation for something that he's witnessed, and his account of what he saw, whatever the final explanation, is convincing.

Rhine had been interested in this case as a possible form of extreme psychokinesis, but at this point it's clear from Rhine's letters and his actions that his interest had begun to wane. Years later, when the movie *The Exorcist* ignited a resurgence of exorcisms, the Reverend Eugene Kennedy would insist to a *New York Times* reporter, "The battle of good and evil is not fought out on the level of demonology. It is much more banal, like in the relations of a husband and wife." The priest could have been speaking for Rhine. If a less paranormal explanation was available, that was the one he'd lean toward. Rhine didn't make any effort to travel to get firsthand knowledge of the case. Still, his curiosity didn't completely disappear. Rhine visited with Schulze the next time he was in Washington, but by then the unexplained events had stopped, without fanfare, unlike the Hollywood ending Blatty gave to it. No one died, although the stressed-out lead exorcist was said to have lost forty pounds during the month-long ordeal.

Perhaps one reason for Rhine's diminished interest in the Maryland case had to do with the fact that the most reported element of Roland's story involved his bed. Poltergeist cases are sometimes open to disturbing sexual interpretations, and beds and bedrooms often figure prominently. It's said that poltergeists are the outward expression of the repressed anxieties and pent-up sexuality of adolescents, usually girls, because poltergeists supposedly appear more often in homes of young girls. Most of the adolescents at the center of the poltergeist cases in the Parapsychology Laboratory archives at Duke University, however, happen to be boys. In this case, Roland's bed would shake whenever he was in it, sometimes to the point of rising up off the floor, and that may have been a little too reminiscent of another story they'd heard involving a pretty little nine-year-old named Bertha Sybert.

Bertha lived in a small, "time-worn" cabin in an area in Virginia known as Wallens Creek. For a time a crowd would gather every night in Bertha's bedroom and watch as she'd climb into bed and quietly lie down. After a while the bed would begin to just barely pulsate, making a noise like a rat gnawing wood. "You could lay your hand on the mattress and feel the vibrations," says Ralph Miner, a witness to events. In a little more time, the mattress started moving up and down, just a little bit at first, but then it would slowly rise higher and higher, until the mattress left the slats, and the headboard made a sound like a piece of wood being scrubbed up and down a washboard. The whole time this was going on, Bertha held very still, her hands at her sides. Four grown men sitting on each corner of the bed couldn't stop the rise and fall. One newspaper reported later that Bertha's sheets would "eerily withdraw from the touch of onlookers." However, two psychologists investigating for Science Service weren't buying it and wrote, "We were amazed at the simplicity of the entire performance," and noted how even neighbors who gathered to watch "voiced skepticism" and were alternately cynical or amused.

Bertha's case was also sadly open to disturbing sexual explanations. Ralph Miner tells the story of a neighbor, a young man, who was visiting with Bertha in her bedroom one night. Bertha was on the bed, and the boy was standing by the window. Bertha complained that something had removed her underclothes. The boy turned and looked out the window. It was a moonlit night and he spotted something white on the ground outside. It was Bertha's underwear on the side of the mountain, among the huge rocks and bushes and trees. No explanation is given as to why the young man was in Bertha's room in the first place or how her underwear ended up in the bushes. Poltergeists, trickery, or a confused adolescent girl acting out would be the best possible explanation. Another might be Joseph Conrad's sobering words, "The belief in a supernatural source of evil is not necessary: men alone are quite capable of every wickedness."

Like poltergeists, there is sometimes an inescapable sexual element to accounts of possession. A demon has taken over someone's body, the participants believe, and there's a liberation of behavior on both sides, because the demon is now responsible for what the child is doing, and for what must be done to expel him. Priests, who have taken vows of celibacy, and therefore are perhaps susceptible to the same repressed anxieties and pent-up sexuality parapsychologists theorized might be responsible for poltergeist activity, have tied young men to beds. Kathleen R. Sands writes in *Demon Possession in Elizabethan England*, "The physical and psychological nature of dispossession and exorcism often lent itself quite obviously to sexual interpretation." She discusses a case of nuns who "manifested their demon possession through the use of dildos both on themselves and each other." Roland would frequently curse and describe sexual acts that at various times included priests and nuns, and Mary, the mother of God. He'd talk about his penis and those of the priests in the room. He'd masturbate in front them, writhe on the bed suggestively, and in addition to the various words that would appear on his body, an arrow once surfaced, pointing to his penis. When a demon is to blame, anything goes. Well-meaning family and priests may feed the rising hysteria. For this and other reasons, exorcists are chosen carefully. They must be sin-free and have led virtuous lives. Most of the unexplained activity in the Maryland case took place at night, when Roland went to sleep. For that reason, the exorcism was always conducted in the evening, with Roland in a bed that sometimes shook and rose off the floor, just like Bertha Sybert's, all of which might have contributed to Rhine's reticence. He needed cases that were free of issues of repressed sexuality. "Rhine saw sexual issues as potentially embarrassing and contaminating one's image in the eyes of the public or authorities and therefore he thought it was necessary to suppress any association to such things," says Jim Carpenter, a psychologist with the Rhine Center.

While Rhine wasn't going to any lengths to pursue the Maryland case, he wanted to be kept in the loop. On May 10, he sent a letter to Dick Darnell, the president of the Society of Parapsychology in Washington, D.C., about Roland. "For the boy's sake, I hope nothing else happens," he wrote, but admitted that for the sake of science he hoped they could get something out of the case before the disturbances stopped. But by the time he was writing those words it was all over. The exorcism had concluded the month before, on April 19, 1949. According to Darnell, Roland's mother was prescribed medicine for an attack of nerves, and arrangements were made for Roland to attend summer school to catch up on all that he had missed. The hospital room where the exorcism had taken place was then sealed for almost thirty years.

Other boys scattered around the country were being plagued by unexplained events that year, and in order to build a hypothesis, Rhine wrote the various priests involved. Objects levitated in broad daylight in the home of a minister in Petersburg, Virginia, where a twelve-year-old boy also lived. And just weeks after Roland's exorcism ended and only sixty miles away, in the small rural town of Lively Grove, Illinois, an eleven-year-old boy was having problems that began with a knocking underneath his bed. Then, as with Roland and Bertha Sybert, sometimes the bed would shudder and levitate, and it couldn't be brought back to the floor even when the boy's brothers and father ran into the room and climbed onto the bed with him. Most of the priests did not write back to Rhine and so he couldn't develop a theory.

Four months after Roland's exorcism ended, Dick Darnell called a special meeting of the Society of Parapsychology and invited the press. After that, stories about the "haunted boy," as Roland was called, appeared everywhere. Rhine opened his own paper one day to read Darnell claiming that Rhine called the case the "most impressive manifestation he has heard of in the poltergeist field." He immediately wrote Darnell that he was not pleased that his name had been tied to the case. "[T]hose who do not like our studies in parapsychology in general will attempt to disqualify us as ghost hunters," he explained. Still, Rhine didn't seem to be too alarmed. The lab would shortly publish a small, restrained piece about the story themselves. Rhine wanted to put out a report on Roland's case as soon as possible, he said, so that people would contact them earlier when cases like this occurred. That would give them a better chance of seeing something for themselves before the episode ended.

More details about the Maryland case have come out over the years. Apparently, it began with a Ouija board; something that even William Fuld, the designer of the Ouija board we know today, didn't believe in. In his application for a patent Fuld wrote, "After a question is asked, the involuntary muscular action of the players, or some other agency, will cause the frame to commence to move across the table." In Roland's case there were those who believed "some other agency" was the devil himself. The scientists at the Parapsychology Laboratory generally tried to steer people away from Ouija boards. When one staff member received a letter from an obsessed Ouija board user, he wrote back that "material gained in such a fashion is not authentic, but is the product of the unconscious just as is that of a nightmare. Therefore may I suggest for your own health that you get rid of the Ouija Board and dismiss the whole thing from your mind." Two-thirds of the people responding to a 2001 survey of Ouija board users said that they had had a negative experience with the board. It sometimes said things that were malicious and frightening. (And yet its popularity continues.)

Sometime after Roland started playing with the Ouija board, the family heard scratching sounds that they thought at first were rats. According to the Catholic Church, possession, which it describes as an attack from within, can begin with outward signs, like the scratching sounds. The outward phase is referred to as diabolical "infestation." Places and houses can be affected by infestation.

Activity in the Maryland case quickly intensified. A diary of the exorcism and the unexplained events that preceded it was kept by one of the priests. A copy was found by workers in the locked room where the exorcism had taken place, just before the building was demolished in 1978. It has since been quoted from in a number of books and articles. Blatty didn't have access to the diary, he says, but nonetheless his book and the movie share many of the same, often lurid details. Blatty embellishes, but a lot of what happens in the book and the movie has a basis in fact, including all the dramatic vomiting. The exorcism lasted a little more than a month, and while the details are sometimes extraordinary, the whole thing quickly gets monotonous in its repetitiveness. The location of the exorcism changed a few times, but the routine didn't. In the evenings the priests would hold Roland down, and follow the instructions laid out in the *Rituale Romanum*, the official book of guidelines on the rite of exorcism. While the priests recited prayers in Latin, there was a lot of spitting, vomiting, and urinating, and pronouncements like "You will die tonight." Scratches would appear on Roland's body and sometimes the marks formed words such as *hell*, *go*, and *spite*. Once, when it was suggested that Roland was missing a little too much school, the words *no school* conveniently emerged. Roland spoke a little Latin here and there, but nothing the priests hadn't already uttered. One priest said he spoke Aramaic. The room got cold, things moved, and Roland was often violent. In the hands of Blatty and director William Friedkin it became a masterpiece of horror. In reality, though, an ordinary explanation can be found for much of what occurred.

The writer Mark Opsasnick researched the story and in 1999 published his findings in an article in *Strange* magazine titled "The Haunted Boy of Cottage City: The Cold Hard Facts Behind the Story That Inspired 'The Exorcist.'" What emerges is a portrait of a lonely boy whose alarming behavior, which supposedly erupted in the beginning of 1949, when Roland was thirteen years old, was actually the adolescent escalation of a long-established pattern of acting out. Opsasnick found schoolmates who described Roland as a bully who was sadistic to both his classmates and animals, and already given to tantrums. Roland's mother is portrayed as obsessively religious and overprotective and the father as nonresponsive. As far as the unexplained feats, Roland "could spit with great accuracy up to ten feet," according to a former neighbor, and "In those days the beds had wire springs and were on wheels and it was not too hard at all to make the bed bounce and move about—it was harder to keep it in one place and his bed was like that."

The only living witness, Father Walter Halloran, who was twenty-seven years old when he assisted at the exorcism, was not convinced that Roland was possessed. When Opsasnick asked him if Roland spoke languages other than English, Halloran said Latin, but "I think he mimicked us."

"Was there any change in the boy's voice?"

"Not really."

Halloran's job at the exorcism was to make sure Roland didn't hurt himself, and he got his nose broken in the process. "When the boy struck you in the nose," Opsasnick asked, "did he exhibit extraordinary strength?" Halloran joked that Roland was no Mike Tyson.

When Opsasnick pressed him for evidence of anything supernatural at all, Halloran said he saw a bottle slide on a dresser when no one was near it, and the bed moved, but he confirmed that the bed was, in fact, on rollers. He also said that he saw the markings on the boy's skin, but when asked if they formed words he said, "It was kind of hard to really tell."

"Was there blood dripping from the marks?"

"It looked more like lipstick."

The author Troy Taylor later came out with a book called *The Devil Came to St. Louis*. Focusing more on the exorcism, Taylor includes accounts from some of the nurses responsible for cleaning up the various fluids that came out of Roland every night, and from people who remembered the screams. Taylor spoke to Father Halloran too, in January 2005. Halloran told him, "I have never been convinced that it fit all of the criteria of a true possession but there was something going on there that I could not explain. For this reason and others, I have withheld judgment on the matter." He didn't think it was a hoax. Perhaps mental illness, but, "I don't know if a mental illness can explain all of it." Halloran died shortly after that, on March 1. He was eighty-three.

Mark Opsasnick's position is clear. "There is simply too much evidence that indicates that as a boy he had serious emotional problems stemming from his home life. There is not one shred of hard evidence to support the notion of demonic possession. . . . Without delving into the dynamics of psychosomatic illness, there is no question there was something wrong with Rob Doe [Opsasnick used this pseudonym] prior to January 1949, something that modern-era psychiatry might have best addressed." Dr. Henry Ansgard Kelly, author of *The Devil, Demonology and Witchcraft*, made an interesting suggestion about one of the stranger symptoms of this case, the words that appeared on Roland's body. Kelly said that the words were a symptom of hysteria, similar to how a hypnotist can raise welts on a subject's skin with just the suggestion they've been exposed to something hot. Gaither Pratt would later ask, "How can we say that these and other similar types of behavior were not the unconscious reaction of the boy himself to the powerful suggestions [from the priests and the family] that evil spirits were at work in his body?" An exorcism works, Dr. Kelly believes, because "what is induced by suggestion can be cured by suggestion."

Troy Taylor believes there is no way of knowing what happened one way or another, but on the whole he is not quite critical enough. Opsasnick makes a more persuasive case; however, Taylor is correct when he points out that there are few occurrences that cannot be reasonably attributed to natural causes. In 1988 Father Halloran told the *Washington Post* that the movie *The Exorcist* "was fair, but 'kind of glitzed it up a bit.'" The problem is, almost everyone who has ever covered this story has done the same—they can't resist making a good story even better. Halloran never believed a demon was responsible for what happened, but objects did move when there was no one there to move them. When you eliminate the sensational elements, which Opsasnick very capably handles in his article, and take only what Halloran and Schulze described that cannot be accounted for, you're left with a troubled kid and what looks like an ordinary and possibly genuine poltergeist case, or a rather dramatic demonstration of psychokinesis.

Today Roland is not willing to talk about the 1949 events, which he claims not to remember. He was abrupt and uncommunicative on the phone with Opsasnick, but he told Taylor that he married, had three children, and now leads a quiet, religious life in the Washington, D.C., area. Halloran called it "a rather ordinary life." Nothing of a supernatural nature has ever happened to him again.

In 1999 the Vatican released a new ninety-page manual for exorcism called *De Exorcismus et Supplicationibus Quibustam* (On Every Kind of Exorcism in Supplication). Among other changes, it bans all media coverage and stresses eliminating possible medical and psychiatric causes for the phenomena. In 2005, when Rome's pontifical academy Regina Apostolorum began offering a course in exorcism, it included ways to tell if someone is genuinely possessed or suffering from psychological problems.

There's been some interesting recent neurological research

in this area. In 2006 investigators at the University of Pennsylvania School of Medicine published results of a test of people who experienced glossolalia, a religious experience referred to as "speaking in tongues." The study came about because of an interest in studying possession states. But the researchers quickly realized it was impossible to find a significant number of cases of possession in time to get anyone into the laboratory, much less getting them to agree to go. Gaither Pratt once pointed out that people who believe they are possessed are in a terrible state and they "want release, not research."

The University of Pennsylvania investigators decided to study glossolalia instead, which has some similarities of experience. Using SPECT imaging (single-photon emission computed tomography), researchers compared images of the brain taken when the people were speaking in tongues with images taken when they sang gospel music. Like someone possessed, "when they're speaking in tongues, they don't feel like they're in charge or control," says Dr. Andrew Newberg, a professor in the Radiology and Psychiatry and Religious Studies departments at the University of Pennsylvania and principal investigator of the study. "When we looked at the frontal lobe, the part of the brain that helps us feel in control, there was lower blood flow and less activity." No changes were seen in the images taken when the subjects were singing. The brain seemed to corroborate what the people who spoke in tongues were claiming—someone else was in charge. "What's causing that to happen," Newberg goes on, "we can't say whether it's God or devil or another part of the brain."

Rhine and Schulze kept in touch for a short while. Schulze wrote that when he saw Roland in the summer he was bigger and that his voice had deepened. "I have an idea that medical examination would show that he reached puberty during this experience." Rhine agreed. "The puberty hypothesis is the one to watch in these cases," he wrote.

At the end of the year, Rhine spoke to an overflowing hall of engineers at the Massachusetts Institute of Technology. They stayed to listen to him from five o'clock that afternoon to seven thirty in the evening. His popularity grew unabated. People would even call Rhine at home in the middle of the night to anxiously relate an experience they had just had. Was it a ghost? A demon? Rhine's son Rob remembers his father listening quietly while the caller told their story, then patiently and gently beginning his answer with "No . . ." When someone wrote the next year about L. Ron Hubbard and Dianetics (known today as Scientology), Rhine wrote back, "Dianetics is not based on any established scientific work so far as I know. It appears to be another cult and doctrine somewhat of the general order of Christian Science without the religious element. However, lots of people need those things and I would not, if I were a dictator, want to rule them out. I would only try to help scientists to find out enough that we could furnish people with something better."

The next spring, the executive committee of the Rockefeller Foundation voted to provide thirty thousand dollars to Duke University for research in parapsychology under the direction of J. B. Rhine. They'd given Kinsey a hundred and twenty thousand dollars in 1946.

By 1950 Rhine had severed his last remaining connection to the Psychology Department when he stopped teaching classes. He would devote himself to research and running the lab. A few years before he had written, "We don't really know what we are." Now he confidently proclaimed, "Instead of bowing before the unexplainable, we begin to experiment with it."

And occasionally Rhine and his colleagues ventured outside

the lab to find it. Gaither Pratt once sat for hours in the twilight on a clear night, a hundred feet from a rain barrel, waiting for it to refill itself with water. The mysterious rain barrel had been in all the papers, and Pratt was sent to Missouri to investigate. As he sat alone in the dark, waiting for the tiny waters to rise, he must have thought, if only fleetingly, "this is it?" The universe opens a crack to reveal one of its many wonders and this is the best it can do? A bunch of pieces of wood lashed together to contain a few swelling gallons of rainwater? Normally he wouldn't have made the trip, but a friend in the area had looked into it beforehand, and while he waited, the water rose. It was a small miracle, but it couldn't be explained. Gaither left for Missouri, but the barrel had stopped performing a week before he got there.

Gaither may not have been too disappointed. If nothing else, the trip was a break in their sometimes monotonous ESP routine, and more than likely he enjoyed the novelty. It was 1950 and news of strange and unexplained events was increasing. There was a paranormal boom, and all around the country people were pronouncing themselves parapsychologists and running around with pens and notebooks conducting field investigations. They'd write up their reports and send them to the lab, but the very serious Dr. J. B. Rhine, who was now fifty-five years old and the undisputed father of parapsychology, could be something of a wet blanket. Although everyone got a hearing, he wasn't exactly open to rain barrels, reincarnation, UFOs, or any of a number of weird events that were currently catching the nation's fancy. They almost always turned out to be a case of fraud, delusion, or wishful thinking, and Rhine had spent a lifetime moving psychical research away from the fringe and into the lab. But the younger scientists were getting anxious. They had found evidence of ESP and psychokinesis. If it existed in the lab that meant it could also be found out there, and sometimes they wanted get to out and see for themselves.

In many ways Rhine's determination to stick to the ESP plan brought them to a virtual standstill, going over the same ground repeatedly. They had yet to develop a theory for ESP. In 1950 they prepared an article based on material from 1938. They'd given ESP tests to fifty mental patients at the Ohio State Hospital. Their diagnoses ran from paranoid dementia praecox to manic-depressive to neurasthenia, but no correlation was found between performance and diagnosis. Worse, no one in the study had performed astoundingly. How many young scientists wanted to recheck twelve-year-old columns of ESP test results when there were things like poltergeists to investigate?

SIX

What they needed were new experiments, and Louie responded by digging deeper into their collection of letters to find them. It was already enormous. Every time an article about the lab appeared, a new batch of letters would pour in and sometimes the response was overwhelming. In 1955, a full year after a *Life* magazine article about the lab written by Aldous Huxley appeared, they were still getting one hundred to three hundred letters a day. While the Rhines were quick to downplay Louie's investigation into the letters as an adjunct to the experimental research, the fact that she was doing it at all was essentially an admission that they were at a standstill and their research had stalled. They had substantiated ESP and psychokinesis over and over, but both the evidence and the abilities people demonstrated were slight. The younger scientists, it turned out, had a point. You needed the lab to establish ESP, but ESP happens out in the world. "Perhaps it was more like astronomy, or geology," Louie later wrote, "in which basic data can only be studied as they occur in nature and not produced by experiment." How could they continue to claim that the "effects trapped so laboriously in the laboratory," as Louie described it, did not exist freely and thrillingly in life? The letters would give them a fresh outlook on the problem.

But for the growing number of people who wrote the Rhines, their experience wasn't just an interesting scientific problem. They needed answers. The writers felt vulnerable in every possible way, and the tone of their letters was frequently defensive and apologetic. "You're going to think I'm crazy, but," they began, and sometimes the senders included pictures of themselves, as if to say, "Look! I don't have two heads. I'm normal." What they wanted was validation and an explanation. What happened to me? Am I alone? Sometimes their reasons for writing were sadder. People with mental illness wrote hoping that what was happening to them could be explained by the paranormal instead. Possession or the voices of the dead trying to communicate might one day end, but the prognosis for schizophrenia was less hopeful. But more than anything else, like John Thomas and Charles Ozanne before them, when someone they loved died and reached out to them, the authors of the letters desperately longed for the learned, rational Duke University scientists to tell them that it was undeniably and incontrovertibly real. Perhaps Louie couldn't tell the people what happened to them, or if it was even real. But with a collection of letters that would grow to more than thirty thousand, she could at least tell them with a voice ringing with the power of truth and authority: You are not alone.

For her first study Louie culled sixteen hundred letters. She wasn't looking for proof of anything. For now, she was just going to see what was there, and to find out if anything could be learned from them. Her main criteria for acceptance was that "they seemed to be sent in good faith by apparently sane individuals." Such loose specifications were bound to invite criticism, but since she wasn't out to prove anything she went ahead. They were just looking for ideas for experiments. The experiments themselves would flush out fraud or delusion.

Louie organized the letters under a number of headings. In the end it came down to three: intuition, dreams, and hallucinations, but it was that last category that got the attention of parapsychologists around the world and others interested in the field. The name alone inspired debate. What others called apparitions, ghosts, or poltergeists, Louie called hallucinations. She called them that because they contained images and/or sounds without any physical basis or external cause. The process of seeing them was therefore mental or "pseudosensory." People objected because the term implied that there's nothing really there. But Louie saw ghosts as an extension of ESP. They were another device for accessing information about the world.

According to Louie, people typically understood ESP as involving two people, the sender and the receiver, with the sender having the more active role. But Louie came to see the process as more "thought-reading rather than thought-sending." There wasn't always an active sender, but there was an active information getter who was using nonsensory means like telepathy or clairvoyance. The idea had been suggested before, but it was not widely accepted, although this is largely how ESP is viewed today. It has evolved into something called remote viewing, which no longer requires an active sender but sometimes simply a solitary information gatherer, acquiring data in ways that remain little understood (and don't involve dead people, presumably).

The crucial thing that Louie picked up from the letters was the fact that people had telepathic experiences regardless of whether an active sender was involved or not. She became one of the first researchers to really push this distinction.

People had a problem with Louie's theory because most believed that when a ghost appeared, the dead person was the one making their afterlife presence known. They are appearing to you. Some suggest that it's a cooperative process. Louie believed that ghosts are brought about entirely by the person seeing them. She was, in fact, saying that when people saw ghosts they were hallucinating.

ESP is an unconscious process, Louie explained. "No one has been able to say: Now I am conscious that a telepathic message is coming to me from origin X." Therefore, if information about a danger came to a person telepathically, there had to be a way of bringing that information to his or her conscious mind. When someone sees a ghost, he or she is essentially taking information gained via ESP and creating a visual drama in order to convey that information from the unconscious to the conscious. For example, a woman wrote about hearing her dead grandmother call to her one night. That caused her to check on her baby, who had fallen to the floor under a heap of clothes. Had she not been awakened by her grandmother's voice the baby would have smothered to death. It was Louie's contention that this woman learned of the danger to her baby via ESP. The voice of her grandmother was simply the vehicle for making that information available. If you were going to construct a ghost to warn you about something, it made sense that you'd evoke the image of someone more likely to want to protect you.

It always came back to the mind for the Rhines. It wasn't so much that they didn't believe in ghosts, it's that they believed in the extraordinary powers of the mind. To them, telepathy and psychokinesis were more plausible explanations than ghosts. But they were cornered by their beliefs. As long as telepathy and psychokinesis were possible, they could never prove survival one way or another, and this would continue to put them in conflict with their backers and other parapsychologists.

The people who disagreed with the Rhines had a point, however. The letters didn't always fit Louie's theory. One told the story of a little girl who was about to jump from one roof to another. Before she scrambled "over a dividing wall which would have caused her to fall into a court five stories below," a man in a blue uniform with brass buttons appeared and stopped her. My name is Bill Johnson, he told her. It turned out that unknown to the little girl, she'd been adopted and her biological father's name was Bill Johnson. Bill Johnson had been a railroad man, which could explain the uniform. Rhine conceded that it was "unlikely that the child would have pulled together the essential information (by ESP) to create a character like Bill Johnson and project it out in front of her as an hallucination." In another case a little boy wrote out an entire message from a dead person in a form of stenography that was no longer in use. Certainly the child couldn't construct a hallucination using an outdated form of communication he'd never learned.

Then there were cases where more than one person saw or heard the same apparition. One woman saw a ball of fire heading toward her son, who was at that moment on a battlefield in Korea. She yelled, "Duck." The son later told her, "Mom, I heard you yell, 'Duck,' and we were sure thankful you yelled because that shell exploded right before where we were standing." Hornell Hart, a sociologist at Duke, pointed out how in one of Louie's examples a total of six people saw the same ghost, including the family dog, which "arose with his hair somewhat bristling, and growled." She'd been calling apparitions subjective, and he asked, "how objective can a subjective hallucination become?"

The most frequent complaint about Louie's hallucination theory was that one would have to possess what came to be called "super-ESP" in order to make her theory work. People who had never displayed psychic abilities would have to "suddenly develop powers of ESP comparable to, if not exceeding, the most remarkable that has ever been experimentally demonstrated," wrote parapsychologist Alan Gauld. He argued that if two people see the same ghost, according to Louie's theory, "two people without any conscious thought of doing any such thing" would have to on "an unconscious level telepathically link up with each other and hammer out the details of an hallucinatory figure which both shall see." In order to explain a ghost that continues to appear in the same spot, "one would have to suppose that someone, not present at the spot, is continually brooding over and inwardly revolving events which once happened there," and then, "somehow persons now occupying or passing through that place become telepathically linked to this distant person, and externalize the information thus gained in the form of hallucinatory figures."

Given how little they'd been able to demonstrate in the lab and how cautious the Rhines themselves were about claims of ESP abilities, Louie's theory is rather far-fetched.

Louie was not dogmatic about survival, however, and her hypothesis was not a casual observation but one arrived at over years of careful study of thousands of letters. "She thought of herself as eminently empirical," her daughter Sally says, "starting out with a wish/belief in survival, but being changed by what she read when she organized cases in terms of who has the most motivation to get the ESP message across to the living—and in the majority of the cases it was the receiver." In the end, Louie knew her explanation didn't cover every case, and on the survival question, she remained open. Three percent of the letters in their collection, she concluded, showed possible evidence of incorporeal personal agency (aka ghosts).

By 1956 Louie had gleaned eight thousand letters whose writers she felt were honest and sane. Of all the things she learned from these letters, perhaps the most interesting discovery was the simple fact that people heard ghosts more than saw them. Footsteps, your dead mother calling your name, doors opening and shutting, knocks, rappings—if people were having hallucinations, they were hearing things more often than they were seeing them, and once again Louie was able to assure the writers that they were most decidely not alone.

"Oh, Mama help me," one woman heard her son call out in anguish. Two days later she learned that her son had been killed on the day she heard his cry. Sounds that ultimately heralded death, like bells or chimes, came up again and again in the letters. One case got the attention of Alfred Hitchcock, who wrote about it for *Coronet* magazine in 1955. A Teaneck, New Jersey, boy named Arne Gandy left high school in 1933 to take a job as a mess boy on a cruise ship. The next year his parents got a letter from him in San Francisco dated January 5, 1934. He told them that they had docked there, but the ship had sailed without him and he was left without his clothes and personal effects. His parents never heard from him again.

On January 10, after his letter arrived, Arne's mother got a phone call at three in the morning. A strange man asked, "Will you forgive your son if he promises not to do it again, and let him come home?" It made no sense, they hadn't been fighting, and she asked him what he was talking about. Then she heard him say to someone else, "When shall we operate, doctor?" Seconds later he came back on the line. "All right, now that he's forgiven, we'll send him home." Who are you, she asked, and the man laughed without humor. He said her son was in a hospital in San Francisco, and he was in bad shape. "But never mind. He's on his way home now." Then she heard an entirely different voice cry, "I am helpless. Here I lie propped up on pillows. I can't move." The voice sighed, groaned, and faded away. When Mrs. Gandy insisted her son be put on the line the connection was broken. Arne's father called the police.

Four months later a body that had been dragged from San Francisco Bay and buried in Potter's Field was identified as the twenty-year-old Arne Gandy. Arne had been in the water for more than forty-eight hours when he was finally pulled out, so he was already dead when his mother received the mysterious call. The cause of death was drowning, but authorities didn't know how it happened. There was nothing to indicate either suicide or slaying, and the last person to see Arne alive was somewhere in Germany. A friend of the family went out to California to identify the body, and while there he arranged to have Arne cremated and sent home. A year later Arne's mother got one more strange phone call. The caller said, "Don't worry about your son. He is fine and in Argentina." The police files for the Arne Gandy case are now missing, not that they would add much. Without leads or witnesses there wasn't a lot the police could do. Arne's case will likely remain a mystery forever. "To her dying day," one of Gandy's nephews says, "my grandmother insisted that the ashes sent to New Jersey were not Arne," although dental records confirmed it. She spent the rest of her life trying to contact him through mediums, which seems to indicate that at the very least she accepted that he was dead, just not completely gone. "Hope in reality is the worst of all evils," Nietzsche said, "because it prolongs the torments of man." And faith trumps science when it tells you what you want to hear.

According to Louie, hearing your son call out to you from a battlefield thousands of miles away, knocks and rappings, and stories like these were all examples of auditory hallucinations. They were sounds conjured by the hearer in order to convey information gained via ESP. It was a satisfying extension of her theory about ghosts and ESP, and Rhine agreed with her conclusion. It fit with everything they had come to understand about ESP, and by continuing to incorporate phenomena that had never been explained before, it seemed to advance their understanding, at least a little bit. However, at the same time that Louie was insisting the sounds were an ESP mirage, one of Rhine's correspondents, one whom Rhine had been a little condescending toward, had nonetheless managed to successfully tape record a "hallucination."

It was a few days before Christmas. Raymond Bayless, a young man who lived in Hollywood, California, and his associate, medium Attila von Szalay, were testing an amplifying system they'd built in order to better hear what they believed were the sounds of the dead. When all the equipment was set up and ready to go, they asked for a voice to wish them a Merry Christmas. It was a paranormal version of testing, testing. They were checking the microphone and speakers before the experiment began. Seconds after they made their request, however, a disembodied voice clearly said, "Merry Christmas and Happy New Year to you all." What they did next was monumental. Apparently only two other people in the world had thought to do this before, and no one in the United States. When they heard a voice where no voice should have been, one of them reached over and turned on a tape recorder.

Raymond Bayless was not a scientist. He wasn't even a college graduate. Bayless was a self-taught artist and illustrator, whose paintings, by the end of his career, were hanging in the National Air and Space Museum, the State Department, and the Pentagon. Raymond was drawn to parapsychology after he saw a sewing machine levitate when he was twelve years old. That was it. A simple household object performing a simple action that objects around the house are not supposed to be able to do. He wanted to find out how it had happened.

It wasn't an accident when Raymond was later drawn to the study of the sounds of the dead. He was always attracted to sound. When he fell in love with Marjorie Brinkman, the pretty girl who would become his wife, he fell in love with her voice first. Marjorie was a soprano who toured as a featured soloist with the Roger Wagner Chorale, which famed conductor Leopold Stokowski called "second to none in the world."

Rhine was always a little leery of the West Coast parapsychologists. "Whoever takes a research position in the L.A. vicinity is going to have worse problems than shoveling snow," he wrote someone who was considering a job in California. "All of the good words, such as 'parapsychology,' have been abused, degrees have been faked, and, in general, a bad odor created to the extent that psychologists are driven to greater defensiveness against anything psychic or occult than can be found anywhere else."

Perhaps Raymond would not have known how to design a lab experiment that could withstand the scrutiny of a country full of academics, but he did come up with surprisingly effective low-tech solutions. He once stuck a piece of fluorescent, glow-in-the-dark tape on a medium's behind. When the lights went out and everyone believed the medium was still in her chair, Raymond was able to watch her as she and the piece of tape moved about the room. "99 percent of what you hear about as being paranormal is either fraud or illusion," he always said. "But that other 1 percent is real." Bayless started sending Rhine reports of his research in the late 1940s. Their relationship was exactly like the one between Rhine and Eileen Garrett. They were always respectful to each other in their letters, but Raymond thought Rhine emphasized the statistical work too much, and Rhine thought Raymond wasted too much time on phenomena that shouldn't be taken seriously.

Raymond and Attila's original plan had been to record the various raps and whistling noises that had been coming from a closet where they'd been conducting experiments. But when they played the tapes back after hearing the discarnate Christmas greeting, there were voices they hadn't heard at the time. Unfortunately, their words were unintelligible, little more than vague whispers. "Nevertheless," Raymond wrote plainly in a letter to the editors of the *Journal of the American Society for Psychical Research*, "the voices were definitely human." They'd repeat the experiment many times, sometimes getting nothing, and at others, a whisper, a word, or a phrase, like the fragile woman's voice who once told them, "It's cold in here."

For some reason, Raymond waited three years before writing the letter to the editor about their experiments. When the parapsychological world finally read the news of what they had done, the response was, inexplicably, silence. But very soon, another man would announce similar findings. Friedrich Jurgenson, a retired Swedish opera singer, composer, filmmaker, and painter, had been out with his wife late one summer afternoon, recording the singing of wild birds. When he played the tape back later, among the chirping birds was the low but distinct voice of a man.

Jurgenson had demonstrated medium-like abilities in the

past, but he wasn't sure that his earlier telepathic messages were coming from the dead. His first guess was outer space. His experiences were taking place during the early days of flying saucers, and now when he and others speculated about the origin of messages from the unseen, in addition to the usual possibilities like fatigue, imagination, schizophrenia, ghosts, and God or other religious figures, the choices now included spacemen. George Hunt Williams, an early "contactee" and author of *The Saucers Speak*, had tried unsuccessfully to tape record invisible voices he was sure were coming from outer space.

Bayless and Jurgenson weren't the first to claim to have recorded invisible voices. Engineer and longtime paranormal researcher Paolo Presi says that in 1952 two Roman Catholic priests, Pellegrino Ernetti and Agostino Gemelli, recorded the voice of Father Gemelli's deceased father. When Gemelli asked the voice if it was really his dad, the voice answered, "Yes I am really your dad, don't you recognize me, my dear *testone?*" "*Testone* is an Italian slang term that means fathead," Presi explained. The church kept quiet about the recording. Later Father Ernetti worked as an exorcist in Venice. An even earlier possible recorder was the Reverend Charles Drayton Thomas, who was said to have made gramophone recordings sometime in the thirties or forties of whispers heard during séances conducted in half-darkness by the medium Mrs. Gladys Osborne Leonard.

When Jurgenson recorded the voice of a man along with the birds he had intended to record, he initially thought that he had somehow captured a stray radio broadcast. But then he went out and tried to record the birds again and this time heard the voice of his dead mother calling, "Friedel, can you hear me? It's mammy." She also added ominously, "Friedrich, you are being watched." In another recording a female voice repeatedly said, "Listen. Please to listen." Three months after Raymond's letter to the editor was published, Jurgenson held a press conference about his discovery.

The world paid a little more attention this time, but not much. The man who ultimately became one of the best known for this kind of research was the psychologist and philosopher Dr. Konstantin Raudive. Raudive came along a few years later, after reading Jurgenson's book. He and Jurgenson became acquainted, and Raudive worked with him for a while before going off on his own. Raudive made thousands of recordings in as scientifically rigorous conditions as he could manage, and in 1971 published a book called *Breakthrough*. He taped voices that begged for acceptance. "The dead live, Konstantin," one voice pleaded. "We are." And from his wife's deceased secretary, who was always skeptical, "Just think! I am!" Raudive got a lot more attention, at least within the parapsychology community, but it was mostly negative. For better or worse, Konstantin Raudive was no J. B. Rhine. He didn't successfully withstand the critical onslaught that followed his book's publication, and recording the voices of the dead does not have the more general acceptance that ESP has today. Although many people have never even heard of the practice, it's still being pursued today, under the name electronic voice phenomena (EVP).

"I'm alive," declares the voice found on the Web site for the American Association of Electronic Voice Phenomena (AA-EVP). The AA-EVP was founded in 1982 by Sarah Estep, a psychical researcher who was drawn to recording the dead because it was something anyone could do, she believed. You don't have "to be a medium or a psychic superstar." The Web site posts links to EVP recordings around the country. One might expect the dead to sound at the very least upset, but the voice announcing "I'm alive," sounds like he's rejoicing. I'm still here. Be not afraid.

"I love you," another voice says. The dead not only remain, they still feel. The woman who uploaded the recording claims it's the voice of her father. A profession of true eternal love should have been heartening, but the words that came across in a low, hoarse, staticky whisper were anything but comforting.

"Help me, David," another voice pleaded. (David made the recording.) The voice asking David for help was hopelessly sad and disturbing. What help could David possibly provide now?

"I'm sad." The weight of the world is in this man's voice.

"Talk to me," another says, imploring and scared.

"I'm so cold."

It's amazing that we don't burst into tears at the sight of a ghost instead of screaming.

"Since there is considerable evidence that experimenters are receiving communications from deceased people," the AA-EVP Web site reads, "this Association maintains a focus on the nature of possible non-physical worlds." But the skeptic James E. Alcock, a professor in the Department of Psychology at Glendon College, York University, believes the activities of EVPers are no different from those of nineteenth-century mediums. "Electronic Voice Phenomena are the products of hope and expectation," he writes. "The claims wither away under the light of scientific scrutiny."

The top three alternative explanations for EVP are:

- Stray radio broadcasts

- Apophenia

- Inadvertent words spoken by the recorders themselves

Most EVPers agree that sometimes they are recording stray radio broadcasts. But not always, especially when the voices address them directly by name. Tom Butler, the current co-director of the AA-EVP, points out that when they get "responses to a question or a comment about the actions of the experimenter," it can hardly be coming from a stray radio broadcast. There have also been successful experiments in which EVP recordings were made in rooms screened for possible radio interference. In 2003 the paratechnologist Alexander MacRae performed an experiment in a room screened according to current U.S. Department of Defense military specifications and TEMPEST standards (TEMPEST refers to U.S. government guidelines for limiting interference from electric or electromagnetic radiation). MacRae "successfully recorded EVP which were later subjected to a listening panel," Butler says. "The listening panel responded to the blind study by correctly identifying the sounds thought to be EVP, and agreeing on the meaning of the sounds."

Apophenia refers to the phenomenon of people finding patterns in sounds where none exist. Studies have shown that people listening to random noise will sometimes impose patterns that aren't there, like words or sentences. In cases where voices have been recorded, different people listening will sometimes hear entirely different words. For instance, if someone played the EVP recording of "I love you" without revealing the content of the message beforehand, skeptics argue that one person might, in fact, hear "I love you," but another might hear "A dove, too," or something otherwise unintelligible.

That EVP recordings are subject to apophenia is not only plausible, but undeniable. The voices on the EVP recordings on the Internet don't always seem to be saying what the recorder believes they are saying. However, Bayless would later point out that though the Watergate tapes could never be transcribed to everyone's satisfaction, no one suggested that Nixon's words, whatever they were, weren't really there. When skeptics argue that the voices are not saying what we think they are saying, that doesn't address the fact that there shouldn't be voices at all.

Skeptics answer that the recordings are in fact white noise that only sounds like voices. However, audio engineers, linguistics experts, and others have been working for decades on speaker identification and evaluation systems, and they know what human speech looks like. A forensics audio examiner was once hired to determine whether a sound from a recording of a fatal accident was a door loudly squeaking or a woman screaming (it was, sadly, a woman screaming). Garrett Husveth, the president of Latent Technologies, who conducts forensic audio analysis for corporate clients and who also records examples of EVP, says, "Forensically, we can prove that they are voices."

The last alternative explanation for the voices on the tapes would be familiar to the staff of the Parapsychology Laboratory. They were sometimes accused of inadvertently speaking aloud which ESP card they were holding, and that's why some of their subjects "guessed" so well. What can the experimenters do but deny it, and point out that the voice on the tape is not their own? (This is actually something that can be either ruled out or confirmed by voice analysis using a device called a spectrograph.)

Tom Butler says, "We can show that people who strongly disbelieve in survival of personality are capable of not seeing or hearing evidence to the contrary," which is reminiscent of the sociologist H. M. Collins's contention that "positive replications by critics are exceptionally rare in science." A determined skeptic can be as unscientific as a determined believer.

"Our working hypothesis for EVP is that the experimenter and/or an interested observer is the channel by which the information is able to be transformed from an etheric aspect of reality to the physical." In other words, unlike AA-EVP's founder Sarah Estep, Butler believes the recorders are something like a medium filtering otherwise unavailable information. "If this hypothesis is correct, one of the things we would expect to see is that the channeling person unconsciously determines what he or she is willing to allow. There is also the characteristic that virtually all EVP (if not all) are in a language the experimenter or an interested witness understands." They tune into only that information which they can process.

Butler has a term for some of the EVP recordings called "echoes of the past." These are "words spoken while the person was still alive, and which are now present in some underlying field of physical reality." Essentially, audio artifacts, which are buried somewhere in an ethereal plane, and are recoverable by a para-archaeologist. For example, Butler refers to a recording made in an old house in the United Kingdom by his colleague David Yee. The house had its drawbridge removed in 1550. "The recording was made in the empty room that was used to control the bridge," Butler says. "In it, you can hear two men and a child speaking what David thinks is an old Latin dialect. Then you hear the bridge being lowered. That is an echo of the past EVP."

Garrett Husveth seems to agree. He rejects the belief that these voices are perhaps earthbound spirits who must be convinced to move on, a common belief about the dead that has been dramatized in books and movies and now every week in the popular television show *Ghost Whisperer*. They don't think you can clear a haunted house or get people to move on. They view this concept as some nineteenth-century throwback, and in the same category as mediums and séances and everything else J. B. Rhine tried to move the field of parapsychology away from. The dead are not frozen in one painful moment from their lives, doomed to live it again and again for all eternity, like a tape stuck on immortal repeat, tragically calling out for help that no one can give. Their souls, if there is such a thing as souls, are gone. But they left a permanent, audible impression.

Gaither Pratt was critical of the early EVP findings and leaned toward psychokinesis as a possible explanation. He thought it was possible that the words were being put on the tape by "PK from the living." Pratt was familiar with the early research of David Ellis, who conducted an EVP experiment in a screened room at the University of Cambridge in the early 1970s. Ellis did, in fact, record voices during his experiment, but he nonetheless concluded that the voices had entered the room acoustically. Someone must have been talking outside the room. After Ellis published, there wasn't much of an effort to perform a more adequately designed and controlled audio experiment. At least once experimenter was unable to replicate EVP, but the study wasn't statistically powerful enough to rule it out either. MacRae's experiments wouldn't come for decades. And without a practical application, government or businesses weren't going to put serious money into EVP. The still, small voices of the dead wouldn't help fight a war or make anyone millions.

Raymond Bayless continued to write over the years, whenever there was a weeping Madonna, unexplained falling rocks, or other strange phenomena. His report about an eerie, golden-amber glowing light that had been appearing at the end of a lonely road near Joplin, Missouri, for more than a century was so thorough it prompted Rhine to suggest that they have a conference in Durham to discuss this particular type of phenomenon.

The light is known in the area as the Spook Light, and has been a local legend for decades, although the exact location has changed. Throughout the years it's been spotted on various stretches of road on the northern edge of the Ozarks, along the Missouri/Oklahoma state line. The source of the light has never been found. The Army Corps of Engineers looked into it during World War II and rather dryly concluded that the Spook Light was a "mysterious light of unknown origin." Most researchers ultimately decide that it's just the reflection from headlights on a nearby highway. But when Raymond wrote his report in 1963, he included evidence of sightings going back to at least the 1800s, years before headlights and highways. In *Ozark Superstitions*, author Vance Randolph also found people who saw the Spook Light "long before there was any such things as a motor car."

Raymond later came out with a book with D. Scott Rogo about a more modern supernatural audio phenomenon called *Phone Calls from the Dead*, which included the Arne Gandy story. Between pornographers and the undead, it's hard to say who finds an application for new technologies first.

Over the years, Louie's observation that auditory hallucinations are more common than visual hallucinations would be confirmed again and again in study after study, although Louie's reports are never mentioned. "We hallucinate sounds more often than sights," Diane Ackerman wrote in her book, *A Natural History of the Senses*, corroborating what Louisa found four decades earlier. "The remarkable thing about the phenomenon is that you generally only hear about it when it crops up in regard to someone who has been diagnosed with schizophrenia," says Daniel Smith, author of a recent book about auditory hallucinations called *Muses, Madmen, and Prophets*. "But if you ask around even casually you come across loads of people who say, 'Oh yeah, I've heard voices my whole life. They're really helpful, not bothersome at all.'" Studies confirm that hearing voices isn't always a symptom of mental illness. Normal people hear voices too. A 1984 study of college students found that "overall, 71 percent of the sample reported some experience with brief, auditory hallucinations of the voice type in wakeful situations." A 2004 report from the University of California called "Hearing Voices: Explanations and Implications" found that "although the majority of people with schizophrenia may hear voices, the vast majority of voice hearers do not have schizophrenia." Freud once wrote in a journal, "During the days when I was living alone in a foreign city—I was a young man at the time—I quite often heard my name suddenly called by an unmistakable and beloved voice." Rhine's friend Carl Jung thought we were hearing our ancestors via the collective unconscious.

Other studies have focused specifically on hearing the dead. A 2004 report, "Auditory Hallucinations Following Near-Death Experiences," found that "among persons who reported having had near-death experiences, 80% also reported subsequent auditory hallucinations." Their attitude toward the hallucinations was "overwhelmingly positive, as contrasted with the overwhelmingly negative attitudes of patients with schizophrenia toward their auditory hallucinations." "Claiming to speak to, and to hear and see the dead spouse (hallucinations) was rather common," researchers wrote in a 1998 Swedish report titled "Hallucinations Following the Loss of a Spouse: Common and Normal Events Among the Elderly." Although not referring solely to auditory hallucinations, forty-seven percent of a group of widows in a 1971 British study said they had some sort of contact with their dead spouse. These findings were confirmed in 1985 by a study called "Hallucinations of Widowhood." In that study, sixty-one percent of the widows reported hallucinatory experiences of their deceased spouse.

Louie said that the people who believe in survival are the most likely to see apparitions. But if you're not schizophrenic and you're hearing voices, what other explanation will suffice? "A lot of people I spoke with who heard voices but were not in any real psychic pain landed on ghosts as an explanation for their voices," author Daniel Smith answers. "Very often, voices seem to carry specific and personally important 'messages.' The obvious next step in the experience is to try and decipher these messages: Who are they coming from? How do they know intimate things about me? Who or what would bother speaking in this manner? Religious figures—angels, God, demons—are one explanation. Another is the dead, often the closely related dead. These days angels come up a lot. But because a belief in the parapsychological seems inextinguishable in all cultures, I suppose it will always be a popular choice when casting about to understand an experience of such obvious force."

Before Raymond Bayless died in 2004, he told his wife, Marjorie, "I will return." They'd been married for forty-five years. "You cannot imagine how much he has returned," Marjorie says. Raymond first appeared five days after he died. "He didn't enter from this world," she said. "He dropped down from somewhere, landed on his feet and started walking." After that Marjorie heard Raymond's footsteps every night for months. "He wouldn't stay long," she said. He'd walk up to her bed, let her know he was there, and then leave after a couple of minutes. "He just wanted to see if I was okay." His nocturnal footsteps were like an audio nightlight, comforting and companionable. He did not abandon her. "He still visits," Marjorie says. Then adds wistfully, "but less and less."

Eileen Garrett once said that "a nucleus of emotional intensity" is what survives when we die. When experiments with her didn't work she thought it was because she wasn't responding to a need, to someone's pain. "I am convinced that without the machinery of emotions and desire on the part of two people, communication has no reason for existing," she once wrote Rhine. "Emotion and love are terrific forces of strength . . . , the only key that unlocks the door to communication between the living and the so-called dead." Raymond came back to Marjorie because she wanted him back.

Rhine continued to keep an eye out for new experiments in the letters they received, discarding those that did not look promising, and recording the voices of the dead wasn't the only area he decided not to pursue. He had been corresponding with a thirty-three-year-old businessman and Wharton School graduate named Morey Bernstein who had visited the lab in 1952. Rhine was so impressed with Bernstein that he asked him to consider coming to work for them. But Bernstein declined, knowing that Rhine had given up on hypnotism as a tool for parapsychological research, something Bernstein was intent on researching. He would later write Rhine, "I could no more change this than you could move away from the parapsychology lab."

A couple of months after Bernstein's first visit to the lab, he began hypnotizing a twenty-seven-year-old Colorado housewife named Virginia Tighe. The world would come to know her as Bridey Murphy. Using hypnotism, Bernstein regressed Virginia back to before she was born. Over the next three sessions, speaking in a brogue, Bridey gave details about her life in nineteenth-century Ireland, and the spirit world she inhabited before she was reborn. It was not the comforting afterlife most people envision, but instead a vague, cold limbo where there were no feelings or pain, the living couldn't hear you, and the dead were distant. "My mother was never with me . . . My father said he saw her, but I didn't." It was a mostly solitary place of waiting where "you couldn't talk to anyone very long . . . they'd go away."

When a friend asked Morey why he wasn't sharing his results with Rhine, Morey wrote, "I do not feel that he will be impressed. This is not the sort of thing which fits into his personal program." Bernstein was right. When a friend hinted that Rhine should participate in Morey's experiments in some capacity, Rhine responded, "We like Morey, but we are not going to be guided by anything as intangible as this seems to be . . . He knows what our standards of evidence are and if he had anything good there is no personal program of mine that would exclude it." Yes and no. If Morey had more compelling evidence, Rhine would have been more interested in the investigation. But had Morey come to him before he started, it's not likely that Rhine would have been open to the experimentation necessary to collect more compelling evidence.

Bernstein's experiments ended after a year, when Virginia became pregnant with her third child. His book *In Search of Bridey Murphy* came out in 1956 and spent twenty-six weeks on the best-seller lists. It was published in thirty languages in thirty countries and started a worldwide Bridey Murphy craze. People held "Come as You Were" costume parties. "My office has been swamped with calls from people who want to be hypnotized," Eileen Garrett wrote Rhine. She would devote almost an entire issue of *Tomorrow* to the phenomenon. An Air Force engineer read Bernstein's book and wrote him, "I couldn't help but think that, after having brought your subject (Bridey) up to Death's door, that it was rather risky to keep going. Personally, I think that if I were doing it I would have retreated at that point because I couldn't be sure what physical effect it might have on the subject."

There was a backlash against the phenomenon from various Christian organizations that had no place for reincarnation in their concept of immortality. When people turned to their church for answers about life after death, the answer the church was prepared to give them was Heaven or Hell, not the return to life on earth. Repeatedly. At a lecture at St. John the Divine in New York the Very Reverend James A. Pike asked parishioners to "contrast this gloomy theory of endless repetition of life (without real continuity or conscious awareness of progress) with the hope enshrined in the Christian faith." But ten years later Pike's oldest son, Jim, shot himself in a New York hotel room and all former afterlife bets were off. Pike went to England and the apartment he had shared with his son before he died. Soon after, Pike, now Bishop Pike, started experiencing a poltergeist. Postcards and books were moved and arranged in patterns, clothes that had been neatly put away were found in disarray, and Pike and the two people who were staying with him were affected by the "more negative, depressive and irrational aspects of Jim," which they believed lingered in the air. They'd wake up with a terrible feeling of anxiety and despair, and Pike would sometimes say awful things while in a blackout state. One of the people staying with him, Maren Bergrud, his mistress, woke up three mornings in a row with some of her hair burned off. On another morning she woke up with painful puncture marks under her nails. Pike met with psychics in London and California and finally had a two-hour televised séance with the medium Arthur Ford in Canada. "I am not in purgatory—but something like hell," Jim communicated, "yet nobody blames me here."

Pike's beliefs had always allowed for life after death, but

not communication with the dead. He changed his mind. "Do you know how exciting it is to come back?" Jim asked during one of their last sessions. "To be dead—but we are not the dead ones, you are the dead ones," Jim said with apparent postsuicidal glee. Pike met with Rhine, who believed the disturbances may have been the result of PK abilities emanating from Pike and not his dead son Jim. Whether from beyond the grave or some more ordinary evil, in their time together a destructive gloom had grown around the small group from the apartment, and it started to pick them off. Maren Bergrud killed herself in Pike's apartment the next year. Her suicide note read, "(a) I am unlovable and (b) you are unloving . . . Maren." Pike dragged her body from his apartment to hers and tore off parts of her suicide note. Soon after, via mediums, Pike began to get confused messages from Maren as well. A few months later Pike moved in with the woman who would become his third wife. The next year, on Pike's birthday, his seventeen-year-old daughter would attempt to kill herself too, and she almost succeeded, but she survived. Pike didn't live much longer after that. The next year he'd die from a fall while doing research in Israel. Scratch the surface of any ghost story and it always turns out more sad than scary.

The scientific community did not accept Morey Bernstein's experiments, of course. "Most human beings are unwilling to face the idea that at death they vanish into husks and the formless ruin of oblivion," the British anthropologist Ashley Montagu wrote in a blistering *Time* magazine editorial. "Not knowing what to do with themselves on a rainy afternoon they, nevertheless, want to live forever." The book "outrages virtually every canon of scientific method, misjudges, mispractices and mistakes the functions and purposes of hypnotism," and "misinterprets facts that have been known to every serious student of psychology for more than a century," while it "virtually overlooks most of the scientific work that has been done on hypnotism during the past fifty years."

Rhine considered Morey a friend, and his public comments were more judicious. "Mr. Bernstein is a capable, likable, well-meaning young man who is having an adventure. He is not posing as a scientist and his book belongs mainly in the field of entertainment . . . I know of no evidence for reincarnation. The best effect of this book is in the way it starts people inquiring and leads to some serious study on the nature of man." It must have been galling to Rhine when in 1958 the AMA decided to recognize hypnosis research as a legitimate field of science. Something associated with reincarnation was getting professional recognition while his painstaking efforts were still officially ignored. Bernstein ultimately agreed with Rhine. "I do not believe that the Bridey Murphy case in itself proves reincarnation. Our case is not strong—in fact it's weak because not enough checks out," he told one reporter. "Yet I believe that the Bridey Murphy phenomenon can be explained only on the basis of reincarnation."

Virginia Tighe, who received threats from religious zealots, would later say, "If I had known what was going to happen I would never have lain down on the couch." Personally, she always kept an open mind about what had happened, and it was said that toward the end of her life she softened even more about the possibility that Bridey was real. Her stepdaughter remembers her saying, "The older I get, the more open I am to it."

Perhaps the most compelling research into reincarnation would be conducted by Ian Stevenson, a professor and the chairman of the Department of Psychiatry at the University of Virginia Medical School. Beginning in the sixties and continuing for the rest of his working life, Stevenson put together his own collection of spontaneous cases, paying particular attention to the cases that suggested reincarnation. Louie had tried to convince him that these cases weren't worth pursing, "But her warning came too late," Stevenson said. Stevenson felt reincarnation was perhaps their best chance to prove survival. "In mediumistic communications we have the problem of proving that someone clearly dead still lives," he wrote. But with reincarnation, they had the easier task of "judging whether someone clearly living once died." Some of Stevenson's most intriguing work was with reincarnation cases involving children, and Carl Sagan would later write "that young children sometimes report the details of a previous life, which upon checking turn out to be accurate and which they could not have known about in any other way than reincarnation," adding that this area merited serious study.

Another craze that eventually reached the lab was UFOs. The modern wave of flying saucer sightings began in 1947, and letters about them started arriving at the lab in the early 1950s, but it was another area that Rhine decided not to pursue. The British historian and philosopher Gerald Heard, whom Rhine had met through Aldous Huxley, formed a group in Los Angeles in 1951 called the Civilian Saucer Investigation. Heard wrote Rhine to say that members of his group "felt that there was an extra sensory element working in this field too," saying the only thing that had a chance of getting Rhine's attention. They were "working in friendly and close cooperation with the new set-up made by the Army into saucer investigation," he explained. "If—as they all say they are now convinced—we are at least in direct touch with an intelligence higher than ourselves, it may well be that it is more advanced in extra sensory perception than we are." Rhine wasn't interested.

A sociologist from Mississippi State College wrote them in 1957 about a sighting. He was also looking for an ESP connection. The response from the lab was polite but not encouraging. Their old friend, the author, psychic, and amateur ESP investigator Harold Sherman, stopped by with a stonelike fragment that he claimed had come from a flying saucer. It had defied chemical analysis, he told them. Rhine was again kind, but not interested in studying the piece of stone any further.

During the fifties, the lab continued in many ways like a glittering salon, where all sorts of ideas were at least considered, even if they were ultimately rejected. A kaleidoscope of visitors streamed in and out. In spite of their professor's warnings, and much to their chagrin, the Parapsychology Laboratory became *the* place to meet some of the most interesting people of their time, from Carl Jung to Jackie Gleason, who called about a show he wanted to do for CBS about psychic phenomena. Gleason was then thirty-seven years old, and the *Jackie Gleason Show* had just debuted on CBS the year before. He was also a serious student of the paranormal and had a library of books about psychic phenomena that would grow to seventeen hundred volumes by the time he died. Gleason seemed to know whom he was dealing with when he wrote, "we are not presenting ghost stories but a true analysis of a mysterious force which is prevalent in the world and very dimly understood." Gleason was always after Eileen Garrett for sittings, and the only reason she finally said yes was because he had called after breaking his leg. But it didn't go very well. Gleason and Uvani apparently didn't have a lot to say to each other and Gleason never called again. He also stopped calling Rhine. "The last I heard about Jackie Gleason," Rhine wrote a friend, "he was on his way to Rome to visit the Pope. I haven't heard anything more about his ESP program for a long time. I think he probably found out that it was pretty complicated." Gleason never did produce a television show about psychic phenomena but gave us *The Honeymooners* instead.

In the fifties, the Nobel and Pulitzer Prize–winning author Pearl S. Buck read Rhine's latest book and wrote him, "I am quite accustomed to the idea of extra sensory perception, from my years in Asia, for people there take it quite for granted." Duke alumnus (Law, 1937) and then Vice President Richard M. Nixon exchanged letters with Rhine for a brief time about political rebellion, civil rights, and the human spirit. Bill Wilson, the co-founder of Alcoholics Anonymous and a longtime correspondent of Rhine's, wrote swearing that the medium Arthur Ford (who would later meet with Bishop Pike) had now been sober three years and Rhine should try to meet him. Rhine had been trying to get Ford to the lab since the early days with McDougall, but as always, something came up every time to prevent it. Rhine assured Wilson that he'd try to meet with him, then joked, "You remember he had an automobile accident on his way to Duke. I do not want him to fall down the elevator shaft if I come to see him."

Aldous Huxley, described by Rhine as "a very gentle person, completely charming and wonderful to know," visited the lab in 1954. Bizarrely, it fell to the brand-new staff member, twenty-three-year-old Rhea White, to take him out to lunch, not Rhine, and she panicked. "I read all his books and I couldn't think of a thing to say to him," White remembers. But she recovered and they had a pleasant enough lunch.

In 1954 the lab lost their Rockefeller grant. (Kinsey's grant would also be terminated that year.) It had looked as if once again Eileen Garrett and Frances Bolton were going to come to the lab and Rhine's rescue, but that didn't end well either. Bolton sent him a check for thirty thousand dollars with a promise to pay the same amount annually for ten years as long as he confronted the survival question. She canceled the grant after one year. Rhine was never going to address the survival issue to her satisfaction and so she put the remaining money into the Parapsychology Foundation, which she had just established with Eileen Garrett.

At the same time the lab lost Bolton, they got their first check from their newest contributor, Alfred P. Sloan, the former president and chairman of General Motors who had established the philanthropic Alfred P. Sloan Foundation and with his friend Dr. Charles Kettering, the Sloan-Kettering Institute. "I believe the question of extra sensory perception, using that term in its broad sense, is more important, in a way, than its impact on the hypothesis of survival," Sloan wrote.

Rhine started putting out feelers about a professional organization he was thinking of establishing; it would eventually become the Parapsychological Association. He saw the association as an important step toward their ultimate goal: membership in and professional recognition from the all-important and august American Association for the Advancement of Science (AAAS). Rhine had given a presentation about their work to the AAAS just before the start of the decade, and a bitter argument had erupted between Rhine and a Harvard mathematician about, as usual, the statistics. While the mathematician conceded that "there is something here which needs explanation and enlightenment," he argued that it was not up to statistics to determine if people had ESP. Rhine envisioned that members of an international group of parapsychologists would help him prove otherwise.

By the summer of 1957 the Parapsychological Association had become an official organization and Rhine had every reason to feel rejuvenated. William Roll was on his way from

Oxford to join the Parapsychology Laboratory staff. Roll was interested in a subject that was getting increased attention across America and now from the lab and Rhine as well: poltergeists. Forget survival and mediums, Rhine said. Mediums "cannot be made into the kind of evidence that the scientific world must have." Poltergeists, on the other hand, were an area that hadn't been completely explored. Perhaps studying them would turn up something more useful. Four poltergeist cases were brought to their attention that summer, "which is a record," Rhine wrote contributor William Perry Bentley. The problem was getting to them before the phenomena ceased.

"I hope you are successful in catching one of those elusive poltergeists," Bentley wrote back. A few months later, they were.

SEVEN

When something strange and frightening happens, people tend to call either a priest or the cops. Given that the events at the Herrmann home in Seaford, Long Island, began with a bottle of holy water and a crucifix, it could have gone either way. On the chilly Monday afternoon of February 3, 1958, Mrs. Lucille Herrmann was at home with her daughter, Lucille, thirteen, and her son, James, twelve, when they started hearing loud popping noises—like champagne corks. They looked around and found something wrong everywhere: A bottle of holy water had spilled in the master bedroom and a crucifix had fallen off the wall onto the floor nearby. In James's bedroom next door, a doll, a plastic model of a ship, and an angel were strewn across the floor, broken. In the bathroom, two bottles were unscrewed and emptied. In the kitchen it was a bottle of starch, and in the basement a bottle of bleach. Mr. Herrmann, forty-three, who was an ex-marine and a veteran of the Battle of Guadalcanal, didn't immediately think "supernatural." His first thought was that the disturbances must have been caused by vibrations of some kind, and he looked around to see if there was any drilling in the area. He also went to Mitchell Field, a local Air Force base (which has since closed), but he couldn't find anything to account for what was happening in his house. Absent any kind of explanation, they did their best to put it behind them.

But on February 6 and 7, when the children were alone, bottles continued to open and spill their contents, including a bottle of bleach in the basement that jumped out of a box and broke on the floor. On Sunday morning, February 9, while the family was in the dining room, the popping and spilling noises resumed. It was the holy water once more, along with bottles of perfume, shampoo, and Kaopectate in the bathroom, paint thinner in the basement, and starch once again in the kitchen. The Herrmanns decided to go to the police.

It's not hard to imagine the expression on the desk officer's face when he took Mrs. Herrmann's call, or the patrolman assigned to investigate. Patrolman James Hughes was out on his regular tour when he was told to head over to 1648 Redwood Path. While he was interviewing the family in the living room, he heard noises coming from the bathroom. Hughes went in and found bottles on their sides, their caps off and their contents spilling on the floor. He put all the bottles in the bathtub and called the desk officer. Later, when Hughes filled out the incident report and came to the box where he would have to classify the crime, he had to think for a moment about what to call it. He settled on "Local Investigation Broken Bottles." They were going to investigate because, if nothing else, a family was actively being harassed. On Tuesday, February 11, Detective Joe Tozzi, thirty-two, caught the case.

Weird news travels fast. A writer for the *New York World*

Telegram had already been and gone, taking all the bottles with him. Tozzi had to track him down in order to retrieve the only evidence he had, which he then sent to the police lab. They weren't looking for ectoplasmic fingerprints, but a natural explanation. James Jr. was at the right age and with a corresponding interest in chemistry, so he was their first suspect. The kid had put something in the bottles that caused the tops to blow. But the only things that were found inside the bottles were exactly what was supposed to be there.

Tozzi's investigation was by the book. He met and exhaustively interviewed every member of the family, and then made a floor plan of the Herrmanns' one-story ranch-style house, diagramming all the activity. His drawings show each disturbance, what moved where, and the position of each family member during each event. Tozzi was an ex-navy guy with close to ten years on the force. His initial feelings were, "This is bullshit. It's the kid." But one day while he was visiting Jimmy at the house a bronze horse hit Tozzi in the back of the legs and fell at his feet. He turned around and no one was there. For the rest of his life he never quite knew how to explain that. He didn't believe in poltergeists or ghosts, but there was no escaping the fact that no one had been there to throw that statue.

By this time the lab had a decent collection of what they were calling spontaneous psi cases. These included accounts of clocks stopping and pictures falling off walls at the moment of someone's death, doors opening and shutting, objects flying off shelves then breaking or exploding, and beds or chairs shaking and leaving the floor. They called these cases spontaneous psi because they felt poltergeists might be not the work of noisy spirits (*poltergeist* is a German word that means "noisy ghost") but an expression of psychokinesis in the field. The problem was, as always, getting to the site of disturbances before they stopped. Hauntings can go on for years, but poltergeist cases typically don't last more than a few months, and they're rare. While the accumulation of cases over the years is impressive, at any one time there aren't a large number of poltergeists or ghosts or haunted houses in America. The few cases the lab had investigated up until this point were also easily explained. One professor Rhine knew was staying in a house in town where he heard strange noises. Gaither spent a night in the house and traced the sound to a wet screen drying out after a storm. Rhine, who had written Mrs. Bolton hoping to be able to tell her that they had found a genuine haunted house, had to give her the bad news. "Dr. Pratt put some screens before the fireplace and he soon had quite a concert of clicking."

When Alfred Hitchcock hired the advertising agency Young and Rubicam to find a haunted house in New York for a party he was throwing in 1956, they couldn't come up with a single viable lead. And this was in a city with hundreds of thousands of potentially haunted apartments. The American Society for Psychical Research hadn't had a request to investigate a ghost for years. An ad was placed in the *New York Times* as a last resort. They got a hundred calls, but most of them were from real estate agents, and not one was about a haunting. Hitchcock finally settled on a "lovely old cobwebby mansion" at 7 East 80th Street, which was "abandoned and gloomy" but ghost-free.

Then a year later, in the space of three months, the Parapsychology Lab was looking into four spontaneous cases. Each involved noises and flying objects and the presence of an adolescent boy or girl. They eventually called these kinds of cases RSPK (recurrent spontaneous psychokinesis) because they continued for at least a short while in one location.

The first was in July, near Hartville, Missouri. Furniture

rose off the floor, and walnuts, jars, pots, and pails of water flew around in and outside of the home of Clinton Ward and his family. The strange turbulence wasn't confined to their house. Once, when they went shopping, objects jumped off the store shelves. Rhine sent Bill Cox, an associate who lived nearby, to act as their representative. The lab had already noted that in all accounts, the Wards' nine-year-old daughter Betty was always there when anything strange happened, and sure enough, on several occasions Cox caught Betty in the act of throwing things herself. It was infuriating because a few times Cox saw things move when he was the closest person to the object and the only one who could have moved it. It led the lab to later write, "scientific detachment requires us to consider the possibility, unlikely as it may seem, that both trickery and psi influence could be involved in the same case."

At the same time little Betty was complicating matters in Missouri, another case was developing in a small town in Oklahoma called Berryhill. In the home of Mr. and Mrs. Cleary Wilkinson and their twelve-year-old adopted daughter, Shirley, plugs were pulled from their sockets, an electric organ was ruined, a sweeper moved through the house pushed by unseen hands, and a pot of water jumped off the stove. Like Mr. Herrmann, Mr. Wilkinson looked for natural explanations. He dug up the water pipes around the house and removed a new fence he had just put in, believing that together they were creating a "magnetic electrical field." The local power company came over three times to check their lines, but nothing was amiss. The family's problems continued. One clock was pushed off a shelf six times, and the motor on their refrigerator was knocked out twice. The second time it happened they decided to stop fixing it and switched to an old-fashioned icebox. "I'm about to crack up," Mrs. Wilkinson told a reporter.

But according to the *Tulsa Tribune*, some of the objects that had flown about were coated with an invisible powder that was later found on Shirley's hands. Shirley then blamed everything on her deceased grandfather, which hardly clarified what was happening. John Freeman, a psychologist who would shortly become a research associate at the lab, went out as a representative for the lab. He saw that nothing moved that wasn't within Shirley's reach. Worse, he observed that "Shirley hallucinated freely with regard to her grandfather." Whether the images of her dead grandfather were ESP-created dramas or symptoms of mental illness he did not say.

The third case that month was out in Clayton, California, near Devil's Mountain, in the home of a retired oil refinery worker named Tony Gomez. There were children in this home too, twelve-year-old Tom and ten-year-old Bob. For two weeks, rocks and pebbles had been falling on their home and the house next door. At one point, all but two windows in the Gomez home were broken. Inside the house, pens, jugs, saltshakers, and ashtrays went flying, tables fell over, and the boys' grandmother had at various times been hit by an onion, a potato, and a one-pound box of salt. They called the police. "I worked for two weeks on this thing, night and day," Constable Vic Chapman said. "We had men on the ridge with 10-power and 15-power field glasses, watching." They were hoping to catch someone throwing rocks, but no one was ever sighted. The two sons were suspected, but the rocks fell even when they were in plain sight. The sheriff's office, which was also investigating, still believed the boys were somehow responsible. But the constable insisted that "no little kids threw those big rocks." The case got a lot of attention on the West Coast, and ultimately led to the formation of the California Society for Psychical Study, Inc.

They looked into the events and pronounced the case genuine. Rhine sent an investigator, but it was over before he got there.

The last case the lab looked into that summer took place in the Rest Haven, Illinois, home of Mr. and Mrs. James Mickulecky while their fourteen-year-old granddaughter, Susan, was visiting. Strange noises were heard throughout the house, beds vibrated, and objects fell around them. The weirdness continued even when they moved to the patio, with all three of them in one bed. When the disturbances followed them across the street to a relative's house they called the sheriff. The sheriff sent his son-in-law. As always, word of the events spread quickly, and a reporter was also there when a shaving brush and a soap dish fell off the shelf in the bathroom, a plaque dropped off the wall in the living room, and pills that were in the reporter's purse somehow left her bag and materialized on one of the living-room chairs. The reporter came back on a different day, and throughout her entire visit there was a muted rapping on the floor by her feet. Everything went back to normal however when Susan's parents came and took her home. No one from the lab investigated this time because it was over by the time they heard about it.

Back at the lab, Rhine and the staff were frustrated by the fact that in three of the four cases, everything stopped before they could get there and one involved trickery. They published their findings, such as they were, in the November 1957 *Parapsychology Bulletin* and ended with this: "If some firsthand evidence could be obtained from observing at least one reasonably clear-cut case—a case that produced the usual puzzling poltergeist effects—further serious studies of these claims as a whole would be in order."

They didn't have to wait long. A few months later, the Seaford, Long Island, case involving the Herrmann family was in all the papers, and a clipping was sent to the lab. After that, articles started showing up at the lab daily, describing recent events in the Herrmann home. It was still an active case. Gaither got in touch with *Newsday* reporter David Kahn first. Everyone was impressed by the intelligence of his accounts, and they believed he was the best person to provide an inside, but less sensational, take on what was going on. Then Gaither. called the Herrmanns directly. It was February 25, 1958. Mr. Herrmann was more than receptive. Things at the Herrmann home were spiraling out of control, and Detective Tozzi was now on twenty-four-hour call. One evening he arrived and found Mrs. Herrmann huddled in the hallway with her children, hysterical. Mr. Herrmann was anxious for the lab's help.

Within hours, Gaither Pratt was shaking hands with David Kahn at La Guardia Airport. Always concerned about being perceived as a ghost hunter, Gaither wanted to avoid publicity in this already highly publicized case. He had proposed that Kahn keep Gaither's presence there quiet at first, and in return Gaither would give his side of the story exclusively to Kahn and *Newsday*. But when Kahn met Gaither at the airport he had to tell him that his editor hadn't gone for the idea and there was going to be a piece about Gaither's arrival on the scene the next day. Kahn would later say he found Dr. Pratt rather stiff.

Mrs. Herrmann met them at the door. She immediately ushered Gaither and Kahn down to the basement where something had just happened. When they got downstairs they found a policeman standing over a broken record player. (Tozzi wasn't available so Sergeant McConnell had come in his place.) The table where the record player had sat was turned over, and records were spilled on the floor. Minutes later Mr. Herrmann, who was a liaison official for Air France, got home from his office in the city. He brought his anxious family upstairs, and Pratt, Kahn, and McConnell remained in the basement to talk.

McConnell said that their plan of having the family call Tozzi whenever something happened was not working. It was over before he got there, and they had learned nothing by showing up after the fact to stare at broken statuary and spilled bottles. The sergeant said he wanted twenty-four-hour surveillance. Then they heard the sound of running feet over their heads. The three men ran upstairs. A lamp had fallen in the master bedroom. Mrs. Herrmann had just left the room herself and no one was there when the lamp fell. Moments later, a bread plate and bread were lying on the dining room floor, thrown by invisible hands.

Pratt was reeling. For once they were not too late. He and Kahn talked together until it was time for everyone to go to bed. Pratt wasn't absolutely convinced yet; nothing had happened where it could be said with absolute certainty that Jimmy wasn't at least nearby. In fact, nothing flew, popped, or spilled when the children were away in school. Still, there was a chance it was genuine.

When Pratt called the next morning Mrs. Herrmann told him the house was packed with reporters. Since his presence there was already known, Pratt felt he might as well talk to them instead of letting them print their take on what his participation meant. But the reporters wanted to know what his findings were before the investigation had really begun. "Even more serious from the scientific point of view," Pratt later wrote, "the presence of so many strangers in the house completely changed the psychological atmosphere. As we would expect, under the circumstances the poltergeist activities ceased."

Gaither left the house for the police station, where he began to explore the extensive and thoroughly unemotional case files of Detective Joe Tozzi. Tozzi had been keeping a meticulous running account of every occurrence in the Herrmann home—every figurine that fell to the floor, the brands of each product that had popped open and emptied (Bright Sail laundry starch and Clorox bleach). He listed every specialist he contacted and what they said and the result of every step taken. In just over two weeks, Tozzi had already amassed a thick stack of single-spaced typed reports. His reports are written without judgment or a hint of humor, even when describing the completely fantastic: "a bottle of shampoo moved along the formica top in a westerly direction and fell to the floor. . . . A heavy glass bowl centerpiece which was in the center of the dining room table left the table and traveled in a north easterly direction and landed about 5 feet away, on the bottom shelf of the corner cabinet."

Tozzi couldn't have been more exhaustive about whom he consulted. At one time or another he had either called, written, or met with the Nassau County Engineers Department; the Long Island Lighting Company, which installed an oscillograph to track tremors and vibrations and checked all outlets and the fuse box; the Brookhaven National Laboratory, which suggested the unusual weather they'd been having (a technical specialist named Robert Zider from Brookhaven Laboratory had shown up with a dowsing rod); the Federal Communications Commission; the building inspector for the town in Hempstead; a tinsmith; the Farmingdale Institute; a physicist at Hofstra University who didn't want to get involved; electricians who checked all the wiring and the TV; Father William McLeod of Saint William the Abbot church, who came on February 17 to bless the house (it didn't help); the Seaford

Fire Department, which checked a well in front of the house; technicians from RCA Communications, Inc., who put up a mobile radio receiving station but explained that radio beams would not have caused these kinds of disturbances in any case; members of the Science Department at Adelphi College; the Nassau County Society of Professional Engineers; Nassau County Water and Power; plumbers; the president of Photo Guard Corporation, about installing a hidden camera; Socony Vacuum; the New York State Water Supply Company; roofers; construction companies that checked the foundation of the house and looked for structural problems; and a medium who stopped by the precinct and offered her services in case all else failed.

Because of all the attention in the press, the Herrmanns had to contend with a continuous stream of visitors and up to seventy-five telephone calls and dozens of letters daily. Tozzi took over the letters, and in his reports he listed each writer's theory about what was causing the disturbances, which included flying saucers, a vacuum in the house, radiation and sunspots, static electricity, a leprechaun, neutrons from Brookhaven Laboratory, someone in the family generating electricity, brain electricity, air and heat pressure, atomic agitation, ultrasonic waves, high frequency skip waves, high-frequency vibrations, heat vibrations, a dead body under the house, friction and gravity, gravitational acceleration, diamagnetic action, magnetism in people, a witch wanting something, a uranium mine under the house, falling tiles from a snowstorm, and Satan.

As far as what to do about the problem, the writers suggested call a priest, call a rabbi, get a dog, give the boy a good licking, burn sulfur to drive the ghosts away, hang a horseshoe over every door, open all the windows for three hours, stay out of airplanes, put money under each corner of the house, read the 23rd Psalm and put all shoes upside down, put a little whiskey in a small glass in each room, put sand on the floor, bless the house, hang a cross at each entrance of the house, try holy water, exorcism, or contact J. B. Rhine (that letter arrived four days before Gaither Pratt).

Not all the letters the family received were friendly. One person wrote to say that Tozzi was a disgrace to the police department, and that the case was a phony and it was all for publicity. Another called Tozzi a no good anti-Semitic wop and implied that he was "a negro."

Arthur J. Woolston Smith, who said he was with the British Insurance Agency, wrote that while he was with the British Intelligence Agency, they had had similar problems on some of their naval bases. A psychic organization had suggested painting all the windowsills at the bases with creosote, which they did and the disturbances stopped. Another person wrote that he wanted to examine the boy, adding that he thought America needed a Democratic president or else there would be another civil war in the South and the Russians would bomb us. A caller from Revere, Massachusetts, said that when she had the same problem she fixed it by installing a chimney cap. Tozzi actually looked into that suggestion, learned about downdrafts, and had a turbine cap installed on the Herrmanns' chimney. It didn't help. He also studied a list of takeoffs and landings at Mitchell Field, but couldn't find a correlation between the airport activity and the events in the Seaford home.

The ghost hunter Hans Holzer called at one point to offer his services. People at the lab always had mixed feelings about Hans Holzer. At Eileen Garrett's suggestion, Holzer had made a career out of investigating ghosts, and he traveled the country with his trance medium, Mrs. Ethel Johnson Meyers, often looking into cases that the lab had passed by. Holzer once wrote Rhine asking for his help with a TV series he was working on. "I need a brief and essentially non-committal letter from you as the head of Duke Laboratory, stating that you are in sympathy with such a TV program." After assuring Rhine that the show would be "rigidly controlled to avoid any fictionalizing or excesses," he ended with, "It is very urgent. Can I not count on your help in this matter?" But Rhine answered that he couldn't make any kind of blanket statement about a program he hadn't seen, and tried to phrase it as kindly as possible, adding that it wasn't that he doubted him, and he wished him well. Eileen wrote, asking Rhine as a friend to please help Holzer, but her letter arrived after Rhine had already turned Holzer down. Nonetheless, Rhine answered, "tell him that with your assurance I would open up more freely and helpfully and we can get together."

Hans Holzer came to the Seaford house with Mrs. Johnson Meyers and said that the problem was their house was built on an Indian burial ground. This was a common explanation from Holzer and his medium. Raymond Bayless once sent Rhine a transcript of a radio show about ghost hunting, and one of the guests refused to continue if Holzer didn't drop "his theories involving vengeful Indians." Later, at the famous "Amityville Horror" house, Holzer and Johnson Meyers would once again hold angry Indians responsible.

Things were briefly quiet at the Herrmann house, and there was nothing more Gaither could accomplish at present, so he left Long Island on March 1 for another field investigation into a dog case in Rhode Island. The lab was always interested in researching parapsychological abilities in animals (they called the work "anpsi" for animal-psi).

But a few days later the newspapers reported a fresh outbreak in the Herrmann home. Just before five p.m. on the day after Gaither left, a dining-room centerpiece crashed to the floor. Ten minutes later a lamp fell in the master bedroom. A couple of hours after that, a globe that had been sitting on the top of a bookcase in Jimmy's bedroom was found sitting on the center of his bed. When Jimmy finally went to sleep his bedroom just erupted. A picture fell off the wall, a brass lamp from his nightstand fell to the floor, and then the nightstand went over too. The kid was cowering under his covers when his father ran into the room. When Tozzi got there it was pandemonium. The entire family was up and Vince Liguori, the Herrmanns' neighbor from across the street, was with them. Jimmy was at the dining room table crying, Lucille was in the kitchen crying, and Mrs. Herrmann was on the verge of hysterics. Tozzi did his best to defuse the situation, then sent Mrs. Herrmann and the children across the street to spend the night at the Liguoris'.

The globe in Jimmy's room had figured in a number of events. One involved the *Newsday* reporter David Kahn. Kahn had been there for several disturbances, things like pictures and lamps and heavy dressers falling, or figurines flying and smashing to pieces. As if things moving by themselves wasn't unnerving enough, Kahn remembers how strangely thunderous it all was. "When glasses fell or were hurled against a dresser they made an explosion—much louder than just a glass hurled against a credenza." But for all that, Kahn had never seen anything move. He was only there to witness the aftermath, but never the flight. Then, on the night of February 24, as he sat alone in an easy chair, the plastic globe from Jimmy's bedroom came bouncing with a quiet thump, thump, thump into the living room. Kahn ran. He went straight to Jimmy's bedroom, but the boy was under the covers. Mrs. Herrmann picked up the globe and said it felt warm. Kahn told Tozzi he honestly didn't think the boy could have thrown the globe and gotten back in bed before he got there.

On March 4, another journalist, John Gold from the London *Evening News*, reportedly stood and watched as a flashbulb he had left on an end table flew slowly and silently across the living room. James Jr. was in the cellar at the time, but directly under where the bulb fell. The flashbulb's small journey would turn out to be only the beginning of a particularly active night. A little while later, they heard a loud bang, but they couldn't find anything out of place. Then, a few minutes after that, three more loud bangs. The raps came from high in the wall, and sounded like they were coming from the kitchen. As they checked the kitchen, there was another loud bang, this time from the bathroom. They began a systematic search of the entire house. The first thing they found was a bottle of bleach in the basement, on its side and on top of a soap powder box. While they were in the basement a crash came from upstairs. Another glass bowl had fallen from the dining room table. They sent James Jr. back downstairs to turn off the radio. A split second after everyone heard the radio click off, they heard the sound of a heavy bookcase on the other side of the basement fall over. They were all quite firm about the timing of the noises. There was no way Jimmy could have turned off the radio and then immediately pushed over the bookcase on the other side of the room.

By March 6 things had gotten so bad Tozzi broke down and called Father McLeod at St. William the Abbot church. Places are as capable of being possessed by demons as people, and early on McLeod and Tozzi had discussed the possibility of an exorcism. At the time Tozzi had decided to hold off; he didn't really believe in exorcism. But now he was ready to try anything. McLeod was out when he called, but a Father Dunne said he would contact Bishop Walter Kellenberg and formally request permission for an exorcism.

Gaither returned to Seaford the next day. This time he brought the newest scientist at the lab, Bill Roll. And instead of telling Kahn they were on the way, Gaither called Detective Tozzi. Tozzi agreed to keep their presence a secret, and for a few days they were able to work quietly at the Herrmanns'. Whenever a reporter showed up, Gaither and Bill would hide in the basement until they left. There weren't a lot of disturbances at the house now, just a "series of loud explosive sounds" without a detectable source.

Then on March 10, Gaither, Roll, and the Herrmann family heard a loud boom beneath their feet. It would turn out to be the very last unexplained event at the Herrmann house. It was just after eight p.m. At the moment the muffled explosion was heard, Jimmy was in the bathroom, Lucille was in her bedroom, and Mrs. Herrmann was in the master bedroom (Mr. Herrmann was away that night). Tozzi was called, then Roll took the upstairs while Gaither took the basement, and the two men began a search. A minute later Gaither called out to Roll. He'd found a bottle of bleach in the basement with its top off, lying on its side. When Tozzi got there, Gaither assured him that no one was in the basement when they heard the explosion, and that he was the first one down there. Tozzi stayed until a quarter of eleven, waiting for something else to happen. But nothing of the kind would ever happen in the Herrmann house again.

LILCO continued to stop by every day to check their oscillograph, but they never found any unusual activity when they examined the tape. Tozzi still consulted whomever he could and reviewed the letters addressed to the family or himself. On March 13 Gaither went back to Duke.

By March 25 Tozzi asked Mrs. Herrmann to take over reading all the mail and forward only ones that contained sound suggestions. He was giving up. He had exhausted all his leads and called everyone he could think of for help. There was nothing more he could do. The bishop did not grant permission to do an exorcism, and in any case, there was no need. It was over. Fifty years later, the case remains unsolved. The Nassau County Police Department dutifully retains Detective Tozzi's aging case files, but presumably there is a statute of limitations for breaking bottles and inexpensive figurines.

Garrett and Bolton's Parapsychology Foundation published their findings first. They actually hadn't investigated while the events were happening and put together their report based on newspaper reports and interviews conducted afterward. In the March-April 1958 issue of the Parapsychology Foundation's newsletter they made their case for the following conclusion. "We have strong enough evidence that the boy might have caused the disturbances by normal means if he so wished. This gives us sufficient grounds to reject the paranormal explanation."

They gave a number of reasons for coming to this conclusion. The only time Jimmy was absent for an event was the time he came home and found his globe on his bed, and he could have put it there himself before he left the house. Mrs. Herrmann was also present for every event, but they didn't suspect her. Instead they point out Jimmy's interest in chemistry and how dry ice could have made the bottles pop, leaving little trace for the police lab to detect, and adding that the police department's "chemical analyses did not cover all the possibilities," in any case. They also point out that bottles that had been painted by the police with an invisible dye in order to capture fingerprints were never disturbed again (the bottles were dusted with fluorescent powder). Their report also referred to Mr. Herrmann's marine-like handling of the children. "In my presence, he commanded both children quite brusquely . . . they were told to come and go, when to speak, and when to remain quiet." "It could be possible that the situation has been so tense for Jimmy that he 'let off steam' by 'pulling his father's leg.'"

Gaither published his forty-five-page report in June. In all, between February 3 and March 10, sixty-seven events took place in the Herrmann home, the majority of which could have been produced by normal means, he conceded. But there was never any direct evidence of a hoax. Tozzi had subjected the boy to a "long and severe grilling" that drove him to tears, but Jimmy continued to insist that he had nothing to do with the disturbances. They had tried to re-create making the bottles pop using dry ice, but the most they could achieve was a slow leak.

Gaither then listed the incidents that were less likely to have been produced by normal means—the times when people actually saw objects move without visible human assistance. On one occasion, Mr. Herrmann saw a bottle of Kaopectate move one way along a Formica top in the bathroom while a bottle of shampoo moved in the other direction. Jimmy was with him, but he was frozen in position at the sink. Another time, their cousin Marie Murtha was watching television with the children at night when a colonial figurine that had been sitting on an end table went flying. Jimmy was sitting on the couch about five feet from it at the time. Marie saw the figurine start to "wiggle, like that of a worm cut in pieces." Then she saw it leave the table. But it happened so suddenly and moved through the air so quickly, it was really more of a blur to her. She would later describe it as a white streak, or a small white feather. The figurine looked like it had turned around and was actually on its way back to the end table when they heard an impossibly loud noise and then it was on the floor. They couldn't tell if it had hit anything, but it wasn't broken. They couldn't explain the noise. Tozzi, Pratt, and Roll all found Ms. Murtha a plausible witness. There was also the time Mr. Herrmann saw an end table in Jimmy's room turn about ninety degrees and then fall over, while Jimmy lay quietly in bed.

Gaither also included all the times the police had the entire family with them in one room while bottles popped and objects moved in other, unoccupied rooms. Gaither himself could verify that he was the first one in the cellar after the last bottle top popped.

Although Gaither's final report was extensive, he made a point of saying that it was not complete. Psychological tests had been administered to the children by members of New York University's Research Center for Mental Health, but the children were also supposed to have been given polygraph tests. Mr. Herrmann kept canceling the appointments, however, giving reasons like family illness or school. He finally refused to allow the polygraph tests altogether.

Mr. Herrmann wasn't happy that Gaither made prominent mention of this, and he responded in the press. Gaither and Roll, he said, were "finding themselves in a corner and trying to get themselves off the hook. A lot of these scientists, so-called, are like a bunch of babies," he went on. "They want to play marbles their way, and if you won't play their way, they pick up their marbles and go home." Lie detector tests were for criminals, he said, and his son had not committed a crime.

Bill Roll said he understood Herrmann's decision. Herrmann had agreed to everything they asked. The children had already undergone psychological testing. "Enough is enough." He was finally drawing a line.

Of the sixty-seven total events, Gaither felt that seventeen could not be explained as pranks, which prompted him to write that "the fraud hypothesis is not supported by the evidence collected by the police, the writers, and other observers." A group hallucination might explain the noises, he said, but not the physical movement of objects. Gaither's final conclusion, however, was that the facts as they stood then did not support reaching a conclusion. The case was unsolved for him too.

It's interesting that Eileen Garrett, who typically gave more credence to stories of this kind, published a report calling the whole thing a hoax while Gaither left open the possibility that this was a genuine poltergeist. Bill Roll thought the fact that the lab had appeared sooner and participated more directly in this astounding case in Garrett's own backyard might have been a factor in her conclusions.

Mr. Herrmann was hurt and angry about the report in Garrett's Parapsychology Foundation newsletter. "Jimmy would have to devote his life to developing mental powers, like an Indian fakir, in order to move even a cigarette box a couple of inches." It was the super-ESP argument, and he has a point.

Gaither couldn't come right out and say it was a poltergeist because it was almost a distinction without meaning. The lab was already leaning toward the hypothesis that poltergeists were actually a manifestation of psychokinesis. A spirit could have been moving the objects, but maybe it was Jimmy, moving them psychokinetically. All Gaither could do was say it wasn't fraud, but that was saying a lot. He was telling the world that what happened couldn't be explained by normal means.

Attention on the Seaford case continued even though the events did not. No one wanted to let go of such a good story. *Life* magazine published an article. Mr. Herrmann appeared on the Edward R. Murrow show, *Person to Person*, on April 11 (and Murrow sent the children a new record player when he learned that theirs had been destroyed by the unexplained forces in their home). Herrmann told Murrow that he was unhappy that officials at the FCC and Mitchell Field, while polite, had kept their distance. He also felt that the fact that the events had involved damage to religious objects ruled out his family. They wouldn't do that.

It wasn't long before Rhine became uncomfortable with the continuing attention to the Seaford case. "I think the fact that Gaither had not shared adequately in the publicity connected with the other work," Rhine wrote his daughter Sally, "may have led him to appreciate this publicity a little too generously. His usual good sense of values on the ultimate payoff was considerably affected by the role of the star actor into which he was thrust almost over night . . . it was obviously exhilarating." But he added, "If Gaither will become a real expert on these spontaneous recurrent psychokinetic effects that will pay off any other debts incurred."

The publicity would increase. In the beginning of October James Herrmann appeared on the television show *To Tell the Truth*, where Kitty Carlisle would ask one of the impostor James Herrmanns, "What is the fellow from Duke University's name?" "Dr. Pratt," he answered correctly. A couple of weeks later, the *Armstrong Circle Theatre* presented a television piece about the events in the Herrmann home titled "The House of the Flying Objects." Gaither appeared along with the family, Patrolman James Hughes, Detective Joe Tozzi, and the dowser Robert Zider. Everyone at the lab was happy with the *Armstrong Circle Theatre* production, except for the participation of the dowser. For the lab, dowsing was something along the lines of table rapping. After the *Armstrong Circle Theatre* show aired, a fresh flood of letters arrived daily at the lab. People with theories and suggestions now started sending their letters to Gaither. Gaither read them all, and put some of them in a box marked "Crack pots. Letters that make no sense." The theories were similar to the ones in earlier letters:

"I would say it is the vagaries of demons."

"There may be an electrical vibration through your family causing a magnetic force—perhaps someone from outer space is trying to prove we have a 'spirit world.'"

"A dead person is buried under your home and until he or she is placed in a cemetery to rest, you will have these goings on."

"Has anyone looked around for money now? If this is an old house money could be planted in a wall."

Their suggestions for combating the problem were also the same, like burning sulfur in every room. "Ghosts don't like sulfur," the writer explained. One child wrote, "I have an idea, why not leave gifts for the spook and say this is for you we are your friends. Try it, it just might work." A few suggested making spice cakes. Spirits "never give up until a young girl bakes them a spice cake and puts it on the back step." One writer urged, "Tear the house down."

While many people shared stories with Gaither that Louie would add to her growing spontaneous case collection, some were from the sadly mentally ill. One woman wrote that she heard voices who told her she had to choose between her son and her parakeet (she chose her son). But some wrote with genuinely useful information. An assistant professor of physics from Upper Iowa University named Philip J. Lorenz sent the results of a study of unexplained events on a farm seven miles west of the Mississippi River in Millville, Iowa. The farm belonged to eighty-three-year-old William Meyer, and it was a familiar story of things moving, falling, and flying. A team of professors in physics and chemistry and six of their students went out to see what they could learn. When they got there they were harassed by a large crowd of onlookers and reporters who shouted obscenities and anti-intellectual remarks and who grew more angry and belligerent by the hour. The house was a mess. Windows were broken and "the interior showed signs of its sudden abandonment and of subsequent damage by the curious." Bedding, clothes, and papers were strewn everywhere. The team had brought along a number of instruments and discovered, among other things, that radioactivity in the yard and basement was fifty percent greater than normal. That wouldn't be unusual, however, "in such a wind-protected hollow, since radioactive gasses emanating from the soil are not readily dispersed," Lorenz wrote. They found something else of interest though. The atmosphere inside the house was initially negatively charged. Lorenz wondered if there might be some correlation between atmospheric electricity and ESP. Using an air ionizer they conducted ESP tests in a normally, a negatively, and a positively charged atmosphere. The best test results, however, were in the normally charged atmosphere. Gaither was intrigued and encouraged them to run more trials, but their study didn't help William Meye, who ultimately abandoned his home. "I was born and raised on that farm," he said, "but I will not die there."

Today David Kahn, who holds the D. Phil in modern history from Oxford and served a year as scholar-in-residence at the National Security Agency, is a respected and conservative journalist who writes about cryptology and military and government intelligence. When asked to look back on the Seaford case he gave the most unexpected answer possible. "My explanation is that the things that happened were caused by a poltergeist. We don't know why or how poltergeists work," he continued, "but similar events have occurred all over the world throughout the centuries. They are not just misperceptions or fakes by people. In other words, poltergeists are a valid phenomenon for which we do not yet have a satisfactory explanation. In all cases that I observed in Seaford, it was always physically possible for the boy to have thrown the globe or flung the glasses that broke or knocked over things. However, he would have had to be almost supernaturally fast to do this. I don't believe he did them. Nor did Joe Tozzi. The point is, that the cause of the poltergeists remains unknown."

James Herrmann Jr. went on to get a degree in engineering, and today is married with a family. He doesn't like to talk about the events of 1958. His sister Lucille however, now a retired schoolteacher, was willing. She doesn't believe in the supernatural. She always thought there would be a physical explanation for what happened in their house, but she is adamant that it happened and that it wasn't caused by her brother or herself. She also had some interesting additions and corrections to existing accounts. She said the contents of the various bottles didn't just spill, they would often disappear, and leave empty, but hot to the touch, containers behind. She confirmed that when things broke they made a "tremendous, huge noise," and once a figurine exploded with such force it broke apart "into the most minutest little pieces," and the flying debris made holes in a mahogany desk like gunshots. Another time, a bottle of ink flew from the dining room, then around a corner and into the front foyer, and then around another corner and into the living room, essentially making two turns, and leaving blue streaks on the wallpaper tracing its path.

Lucille and her brother remember Detective Tozzi with a great deal of affection. "He protected us." When things went particularly crazy, Tozzi would come running to comfort the scared children, sometimes in the middle of the night. "He was very sympathetic to us. He'd stop by sometimes and have a cup of coffee with my mother, to see that she was okay. He was very kind." Gaither Pratt and Bill Roll, on the other hand, responded by giving the children ESP card tests. It seems almost comically insensitive now, but it was the only method they had to see if the kids had abilities that could explain the disturbances. Pratt and Roll were helping in the only way they knew how, but it's not surprising that Lucille and her brother have such fond memories of Tozzi and rarely think about the Duke scientists at all.

Not long before the *Armstrong Circle Theatre* production aired, Rhine asked Pratt to put the Seaford case aside and move on. No one at the lab was happy with his decision. "The Coffee Hour at the lab today was the scene of unexpected fireworks, when Dad tried to package up the Seaford Case," Louie wrote her daughter Sally. "After all the publicity the case has already received, now the TV people want to make it the subject of one of their shows, but Dad said enough time and energy from the lab has already been expended on the case. After all, it's not an experimental issue . . . I thought he explained why he took this stand very well, but one member of the staff exploded and I sensed the rest were with her."

Rhine wrote Sally to explain too. "The spectacle of seeing us make too much of cases that cannot be firmly identified as justifying scientific interest is not one we want to encourage and extend." Sometimes the very characteristics that made Rhine so right for their battle for acceptance had a steamroller effect on the people around him. Everyone in the lab was excited by the Seaford events and wanted to continue the investigation. But "being patient and supportive doesn't go hand and hand with being tough and resilient," Stanley Krippner, a former colleague pointed out, and Rhine's concern, it turns out, was warranted. Just a couple of years before, Martin Gardner had written Hubert Pearce to ask him if he wanted to come clean and admit that he had cheated all those years ago. Twenty years of answering no was still not sufficient and the acceptance battle was not yet won. The Seaford case was not building on the results of previous experiments or investigations. It was time to move on.

Unfortunately, Rhine's single-minded dedication didn't satisfy his critics and not only frustrated his staff, it alienated some of the people financing the research. The next year Charles Ozanne took all his money from Duke and the lab and put it into the what he called the Survival Research Project. No more ESP cards; he wanted research more directly connected to the question of life after death. Shortly after that, Rhine would lose the Alfred P. Sloan contributions as well.

Still the public's fascination with their work grew unabated. The experiments of the Duke Parapsychology Laboratory continued to filter out into art and popular culture. A main character in Shirley Jackson's novel *The Haunting of Hill House*, Theodora, is portrayed as a Hubert Pearce ESP card-guessing star. "The name of Theodora shone in the records of the laboratory," one passage reads. But in the next sentence Jackson writes, "perhaps the wakened knowledge in Theodora which told her the names of symbols on cards held out of sight urged her on her way toward Hill House," implying greater ESP abilities than had been so far demonstrated. It was a leap into magical ESP territory, indicating that Shirley Jackson and the general public didn't really understand what ESP was or what it could do, a misunderstanding that would soon have tragic consequences.

EIGHT

On July 13, 1960, six-year-old Bruce Kremen became the fourth child in four years to go missing in the Angeles National Forest in California. He'd been on a camping trip with eighty other children in the heavily wooded area of Buckhorn Flats when he disappeared. He was last seen wearing a white T-shirt that read "YMCA Summer Fun Club," and his last words were "I want to go home." Hundreds of people searched the area for two weeks. They used helicopters equipped with loudspeakers: "Bruce, if you can hear me, come out into the open. Wave your shirt so we can see you." Dogs and mounted units were brought in along with dozens of groups of mountaineers and hikers. Women from one volunteer group said they were feeding up to one hundred fifty searchers daily. But five days later he was still missing. When the missing person flyers went up, someone handwrote "ears stand away" beside his picture, and you can see the stick-out ears on either side of his sweet, smiling face. On the sixth day of the search veteran forest rangers conceded that there was now only a slim chance that Bruce would be found alive. Three days later if he was still alive, Bruce Kremen turned seven. His parents appeared on television and the story was printed in newspapers throughout the country, but it didn't help. He was still missing four months later when his father, Joseph Kremen, wrote J. B. Rhine. "Please, I beg you, if you yourself can help through ESP or can call on someone who can help us locate our little boy, please contact us."

This put Rhine, a father himself, in an impossible position. "You may be sure I would do anything I could to help you and Mrs. Kremen in this terribly distressing situation," he began. But he didn't want to hold out false hope. "We do not know enough about the abilities we are studying to be able to apply them reliably, and those who claim to do so have not impressed us as having a sufficient basis for such claims. . . . The worst part of it is that there is no adequate assurance that the impressions that come to the mind are due to ESP and are reliable even when they actually are." But Rhine was reluctant to remove what the family may have felt was their last shot at bringing their son home, and so he gave them three people to contact, his old friend Harold Sherman who lived in California, a psychic in Dallas, and Peter Hurkos, a former house painter from the Netherlands who acquired his abilities following a fall from a ladder and who was now living in Miami. "But you would have to watch that you were not caught up in a demand for a fabulous fee that you could not afford," he cautioned Kremen. Rhine didn't expect this step to help and he wanted to make sure it didn't hurt.

Bruce's parents went to Hollywood to see Harold Sherman on November 22. Their need was so great and distress so palpable Harold had a hard time working in their presence. "The mother wants me to tell her that her boy is alive and will return to her," he wrote Rhine. The only impression Sherman got that day was of a parking lot and a highway by the trail where Bruce had been walking. Sherman told the Kremens that he needed something the boy had worn on this trip. Bruce's parents didn't have anything like that, but they said they'd return with a favorite hat.

By the time Harold wrote again in December, Mrs. Kremen was on the verge of a nervous breakdown. Their son was still missing, and when they came back with Bruce's hat, Harold took hold of it and saw Bruce die. A man in a red-checkered shirt and khaki pants had convinced Bruce to get in his car. They drove for a mile and a half before turning off the road into a dry gully between two hills. "I can't write down what I seem to see happening," Sherman wrote. Instead he skipped to a description of the aftermath. He saw beer cans in the station wagon the man drove, and a drop-off in the dense underbrush where he saw Bruce's body and clothes. Sherman said Bruce's killer had a dark complexion and was perhaps Mexican. He was around thirty-eight to forty years old, married, had two or three daughters, and he might have a record of sex offenses against boys. "I feel this compulsion will lead him to frequent areas where other groups of boys and young people are gathered—that he is now brooding and planning to return to the Buckhorn area next year, feeling that he has gotten away with his crime, and strongly tempted to vicariously re-live this experience by re-visiting the scene." That particular forest "is a favorite haunt with him."

Like Rhine, Sherman was in a no-win position. He didn't want to tell the Kremens that their son was dead—"it hurts me unspeakably to have to make such a report," he wrote. Besides, he could be completely wrong, he told the Kremens. He also said that he was sure there was life after death, and if Bruce had gone on, "he is alright now." Another psychic in a similar situation would write to the loved ones of a murder victim, "Don't agonize over images of the horror and pain of that death . . . the soul leaves immediately . . . death is not terrible even in these situations." (There is medical evidence to support her claim. When discussing the case of a murdered child, Dr. Sherwin B. Nuland wrote in his book *How We Die*, "It is not farfetched to believe that the human body itself knows how to make these morphine like substances [endorphins] and knows how to time their release to correspond with the instant of need . . . I am convinced that nature stepped in, as it so often does, and provided exactly the right spoonful of medicine to give a measure of tranquility to a dying child.")

In January Mr. Kremen sat down and wrote out a six-page letter to Rhine. His son was now missing six months. In Kremen's view, Sherman and the psychic in Dallas hadn't given him anything concrete, and so he was casting them aside. Instead he was pinning all his hopes on the last psychic on Rhine's list, Peter Hurkos. It turns out Kremen had already called Hurkos in September, two months before he had written Rhine. But it had been a busy time for Hurkos. The television show *One Step Beyond* had aired a two-part feature about him earlier in the year, and Norma Lee Browning of the *Chicago Tribune* was working on a feature that would come out later in September. Hurkos didn't get back to Mr. Kremen until late November, when Hurkos's lawyer called and said send money and we will come. Kremen sent money, and his letter described what happened next.

Hurkos showed up in California on Monday evening, December 5, 1960. The next day he and the Kremens drove to the campground. At a certain spot Hurkos said he saw a man and a woman pick Bruce up. Kremen wrote, "By this time I could see him in a cold sweat. Hurkos then turned to his lawyer and said, 'Did you ever know me to be wrong when I sweat like this.' And he was wet," Kremen wrote hopefully.

The next day Hurkos and Kremen drove up the coastline. The day after that Hurkos studied maps. On Thursday night Hurkos announced that Bruce had been picked up by a childless couple in their forties who lived on a farm in a small town in Oregon. And that was it. His job was done. As far as Hurkos was concerned, he had given Kremen what he paid for, it was up to the police now, and he was leaving the next day for another appointment. The Kremens were hysterical. They knew this information was too vague. A small town in Oregon? A nameless, faceless couple in their forties? Where would the police even begin? They went to Hurkos's hotel room the next morning and begged him not to leave. Hurkos promised them that he'd go to Oregon himself to find Bruce. Four days later he called. He was in Oregon, he said, and about to begin his search. He said he was getting the name "Roy" or "Rossi." They waited. Days passed. A week went by and the Kremens still hadn't heard from him. They finally called his wife in Miami and found out that Hurkos was not in Oregon, but there with her. He had gotten sick and had to leave, she said. He was going to be in Los Angeles in mid-January and he would contact them then. The Kremens had no choice but to wait. It was now a week before Christmas. Bruce had been missing for more than five months.

The holidays passed. Then mid-January went by without a word. Mr. Kremen called Mrs. Hurkos again. Hurkos had been in Los Angeles for two weeks, she said, but he'll call. He didn't call. "Of course my wife clings to his impressions and even I want desperately to believe that this could all be true," Kremen wrote Rhine, "but I wonder what you make of all this and would appreciate an answer."

Rhine wrote back and copied Hurkos. The letter is addressed to Mr. Kremen, but the whole thing reads as a message to Hurkos. Rhine never openly accused Hurkos of anything, he was very diplomatic, but his meaning was clear: You are tormenting this family and you must stop. "I do not think any decent living human being could be so cruel as to break off and leave you in suspense if he could help himself," he wrote. Which is of course exactly what Hurkos had done and Rhine was calling him on it. Rhine explained to the Kremens that this was not an exact science, and that no one with this ability, if they in fact possessed it at all, could control it, so be patient. But the underlying message runs all through his letter. "What better use in all the world could he find for his ability than trying to help you find the little lad you have lost?" Rhine concluded his letter with a final warning to Hurkos, "I cannot believe he will give up your case entirely—not if he can really do anything, and it ought not to be a question of money . . . his expenses and a reasonable fee for his time, but he ought to want to do this because, if he has, as he believes, a gift that has been given him, he cannot think of anything finer and more wonderful to use it for than to help you in this very deep distress you are in. Shame on him if he doesn't, I would say, and I think all the world would feel the same."

There is no record of any further communication between Hurkos and the Kremen family, although it could be that the records are lost. His widow (not his wife at the time of the Kremen disappearance, but a later spouse) suggested that it had never been Hurkos at all, but an impostor. However, given that Rhine had the correct contact information for Peter, this explanation isn't likely.

Bruce Kremen was never found. Mrs. Kremen, who is now eighty-five years old and a widow, becomes enraged at the very mention of Peter Hurkos. Her anger overcomes her and she can't even bring herself to explain. She talks about Bruce instead. She knows he is dead, she says. For almost fifty years she has tortured herself with self-recrimination. The agony in her voice as she explains her reasoning is as strong as if this all happened yesterday. If only she had said no to moving the family from Brooklyn to California, she argues, Bruce would still be alive. "My life went down the toilet after that," she says. When the murderer brutally stopped an unhappy little boy from making his way back home he also effectively destroyed the home he was trying so hard to come back to. Peter Hurkos's fumbling insensitivity made an unspeakably horrible situation worse.

When the Missing Persons Unit was called to see where Bruce's case was left all those years ago, after looking for the file they came back on the phone and said, "The Cold Case Squad has it." There are thousands of unsolved homicides in every major city, and a cold case detective is not going to pick up just any case without a reason, especially a missing person case, which is the most notoriously difficult to solve. If a detective drove over in order to sign out an almost fifty-year-old case this could mean only one thing—they had a viable lead.

It had come from author Weston DeWalt, who found new evidence while researching Tommy Bowman, another boy who went missing in the same general area three years before Bruce. The evidence linked Tommy's case to Mack Ray Edwards, who had confessed in 1970 to sexually abusing and killing six children, including eight-year-old Stella Darlene Nolan. Edwards described to law enforcement how he had raped and strangled Stella, then threw her off a bridge in the same forest where Bruce disappeared. She was still breathing when he came back later, so he stabbed her to death and then buried her along the Santa Ana Freeway. Her skeleton was dug up seventeen years later, after Edwards told police where they could find her. Edwards was eventually convicted of murdering three children, and while in prison he confessed to killing eighteen more, but before he could be questioned further, he hung himself with a television cord in 1971. Unlike most murderers, who rarely express genuine remorse, Edwards begged to be put to death, and he killed himself because the state wouldn't do it fast enough. Ultimately it was his last self-centered act. He left this life before providing answers for the families of the rest of his victims.

According to Weston DeWalt, who still actively investigates Edwards, a number of California law enforcement officials believe several unsolved child disappearances and homicides from the 1950s and '60s may be attributable to Edwards. At least six of those cases are being reexamined in light of newly gathered information. "I think it highly likely that Bruce was one of his victims," DeWalt writes, "and is buried at the site of a road construction project upon which Edwards was working at the time." As of this writing, Detective Vivian Flores has transferred from the Los Angeles Police Department Cold Case Unit and is now the Van Nuys Detective Gang Unit supervisor. She took Bruce's case with her, however, along with the case of another missing child that is connected to Edwards, Roger Dale Madison, and continues to do whatever she can.

Weston DeWalt agreed to review both Harold Sherman's and Peter Hurkos's descriptions of Kremen's abductors. Harold Sherman said Bruce was dead (Hurkos said he was alive) and that his killer was thirty-eight to forty, had a dark complexion, two or three daughters, and a history of sexually abusing boys. Sherman also said the killer was stalking the area for more victims. At the time of Bruce's disappearance, Mack Ray Edwards was forty-one. DeWalt responds to the rest. "While he did not have a dark complexion, he was a heavy equipment operator in Southern California and often appeared deeply tanned. In 1960, he was married and had two adopted children, a girl and a boy—both under the age of 10. I have evidence that he was molesting girls as early as 1946." DeWalt has also established "that in his off-hours from work he often spent a lot of his time 'driving around.'" He and the detectives think that Edwards was cruising for victims and revisiting the scenes of his crimes and body dumps. Detective Vivian Flores added that "the first psychic [Sherman] has some similarities regarding the shirt and pants." Sherman had said a man with a dark complexion dressed in a red-checkered shirt and khaki pants had convinced Bruce to get into his car, and back in the 60s, two possible witnesses had described a "deeply tanned" man in khaki pants and red-checkered shirt following behind a boy believed to be another of Edwards's victims.

Hurkos came up with the names "Roy" and "Rossi," and Edwards's middle name was Ray. That's about it. He said that Bruce had been picked up by a childless couple in their forties who lived on a farm in a small town in Oregon. DeWalt was blunt. "Nothing I have uncovered in my research would suggest any such scenario." Detective Flores agreed that Hurkos did not come up with anything remotely significant.

There are many who believe that Peter Hurkos had a gift, and it could be that Hurkos did in fact try his best to help the Kremens, and therefore can't be blamed for failing. But he can be held accountable for how he cruelly left the family hanging and strung them along, making promises and pronouncements when he should have been more circumspect. This was a little boy and not a pack of ESP cards. Hurkos had a history of exaggerating his accomplishments. He claimed to have helped Scotland Yard solve several murders for instance, but when they were contacted in 1961 by someone investigating Hurkos, the investigator couldn't find anyone who had ever heard of him. The commissioner denied that they ever had anything to do with him. Perhaps Scotland Yard simply didn't want the fact he worked with them known.

The Miami police always denied working with him. "We prefer to stick to accepted police investigative methods," one police chief said when Hurkos went down there in 1957 to help solve the 1954 murder of six-year-old Judith Ann Roberts. "We haven't resorted to using crystal balls yet. I certainly haven't engaged him and I have nothing to do with him." But this was a huge case at the time. A little girl was taken one night from her grandparents' home and found just before daybreak on an abandoned road under a group of mangrove trees, savagely beaten, strangled with her nightgown, and raped with a tree branch. It's not surprising that detectives were willing at least to hear Hurkos out. It would be almost immoral not to. Irving Whitman, the original detective on the case, who is now eighty-six, confirms that he spoke to Hurkos. He remembers him as a perfectly nice guy, but he didn't impress Whitman and he didn't provide any useful insights into the case. Whitman missed the show Hurkos put on at a local church. A bunch of objects had been placed on a stage. "Hurkos, sweating and apparently extremely tense, picked up a bag containing a child's shoes," a local reporter stated. Judith's grandparents had provided the shoes, and they informed the crowd that Peter correctly picked out their granddaughter's shoes. It didn't help find the murderer, however.

Hurkos said he'd crack Judith's case within two weeks, but fifty years later Judith Ann Roberts's murder is still unsolved and under investigation by Detective Andy Arostegui of the Miami Cold Case Squad.

The month before Bruce Kremen went missing, Hurkos led Virginia police to the wrong suspect in the murder of the Jackson family. Hurkos had been hired by a local psychiatrist named Riesenman who had two patients among the suspects. Peter claimed he could "pick up brain waves" that lingered in crime scenes and demanded to be taken to where the bodies had been buried. Again, what choice did the police have? They had four particularly horrible unsolved murders —the baby of the family had been buried alive—and a town full of frightened citizens. They took Hurkos to the two graves where the killer had buried the family. "I'm a very sensitive man," Hurkos told local reporters. "I pick up signals just like a radio. What I pick up is subconscious." He was going to solve the case, he assured everyone.

On June 8 Hurkos announced that he knew who the killer was. His description matched John A. Tarmon, a trash collector the police had questioned and released the month before. The police went to Tarmon's home with a warrant, but found nothing to justify arresting him. The next day, Dr. Riesenman and Hurkos visited Mrs. Tarmon. This time she admitted that her husband had acted "funny" around the time of the family's disappearance. The police went back out and picked Tarmon up, and a doctor convinced Tarmon's wife to sign a commitment petition. At one fifteen in the morning, Fairfax County Judge J. Mason Grove got out of bed and convened a lunacy hearing. Tarmon was committed for an indefinite period to a hospital roughly three hundred miles away, and Peter Hurkos left Virginia after assuring everyone that Tarmon would soon confess.

Three days later an editorial was broadcast on WTOP television and WTOP radio. "The spectacle of Peter Hurkos roaming the Washington area trapping faint old 'brain waves' would be mildly funny if, in the process, those who believe in what are described as his 'supernatural powers' were not threatening to tread heavily on some fundamental civil rights." Rhine told WTOP News that he had never found anyone with abilities "remotely approaching" the ones claimed by Hurkos. The ACLU quickly intervened. A week later Tarmon was out, and on June 25, the FBI arrested the real killer, Melvin D. Rees.

Years later Professor Walter Rowe, from the Department of Forensic Sciences at George Washington University, looked into Peter Hurkos's work on the Jackson family case. "Most of his 'amazing' hits involved information readily available in the newspapers."

It's certainly not Hurkos's fault that the police arrested the wrong man, or that the authorities committed him in the middle of the night. But even though Hurkos had the wrong man, *Chicago Tribune* reporter Norma Lee Browning later called this "one of Peter's most fantastically successful cases." Peter's mistake, she said, was actually "an intriguing case of psychic double identity, or crisscrossed vibrations." According to Hurkos, Melvin Rees and John Tarmon were friends, and Rees had lived in the house that Tarmon was living in when he was arrested. Although Tarmon had been living in a house fitting Peter's description when he was initially suspected, he wasn't living there when he was arrested. Nor was this the house Peter had visited, but apparently Peter didn't notice. Virginia police today can find nothing to substantiate that Rees ever lived at the earlier address.

Hurkos would make a similar claim of mistaken psychic identity when he went to Boston a few years later to help solve the Boston Strangler murders. After coming up with some details about the crimes that admittedly were known only to the police, Hurkos pointed them to a fifty-seven-year-old shoe salesman named Daniel Moran who also had been a suspect at one time. Like John Tarmon, Moran was committed to a mental institution following Hurkos's announcement. This time Peter said that Daniel Moran and Albert DeSalvo, the man who was ultimately convicted of the crimes, had "roommated together in the same mental hospital. That is how DeSalvo knows everything to confess." But DeSalvo and Moran were never roommates, they weren't even in the same hospital. DeSalvo was in Bridgewater State Hospital at the time and Moran was in Massachusetts Mental Health Center. Although there have always been serious questions about DeSalvo's guilt in some or all of the killings, there has never been any evidence tying Moran to any of the murders. That doesn't stop the official Peter Hurkos Web site from proclaiming his Boston Strangler work as one "of his most illustrious cases."

Hurkos managed to insinuate himself into many famous cases. When Michael Rockefeller, the youngest son of then New York governor Nelson Rockefeller, vanished while on an anthropological expedition in New Guinea in 1961, Michael's older brother Steven flew to Wisconsin to meet with Hurkos. The most prevalent theories at the time were that Michael had drowned, was killed by crocodiles or sharks, or was eaten by cannibals. Peter told Steven some things about his wife which were true, and then said that Michael wasn't dead. "He is with native people in a cave-like dwelling. He rested (either he slept or was unconscious from exhaustion) for several hours in the swamp after reaching land. When he started swimming with the oil drums, he drifted downstream for 16 or 17 kilometers. He was not bothered by sharks or crocodiles. After resting in the swamp, he realized he couldn't stay there, so he went straight ahead through the swamp to higher land. He didn't reach the mountains, but is in a flat area where there are open fields and woods. Possibly the natives picked him up while he was still unconscious and took him to their mud dwellings." He's alive and "probably alright," the president of the recently established Peter Hurkos and Associates Foundation, Inc., wrote. "If you do find it necessary to go out there, I should strongly recommend that you take Peter with you." Hurkos later said that Michael was making no attempt to return to civilization, in order to punish his parents for their impending divorce. Hurkos's foundation would close within a year. Michael Rockefeller was never found.

In 1969 Hurkos was hired by friends of Jay Sebring, one of the victims of the infamous Manson Family murders. Roman Polanski, whose wife Sharon Tate was also killed that night, was skeptical, but he allowed Hurkos to visit the scene of the crime. Hurkos told the press that the killers were three men and one woman who were friends of Sharon Tate's and who had been turned into "frenzied homicidal maniacs" by LSD. The brutal murders were part of a black magic ritual called "goona goona." In reality, the killers were three women and one man, and all were strangers to Tate. However, according to Browning, Hurkos named four people, and one name was correct: Charlie. (There were two Charlies connected to these murders; Charles Manson and Charles "Tex" Watson.) Hurkos also accurately described Charlie as a little man with a beard who lived in a mountain area with hippies, though in 1969 America everyone was blaming hippies for anything that went wrong in the country, especially those in California who liked to live in the hills. But if Browning is to be believed, this is one for Hurkos.

Still, Professor Walter Rowe called Hurkos's technique "a pastiche of common sense, stereotypes, and popular mythology." If you throw enough out there that might fit the given situation you're bound to get something right. "Psychics often speak in a stream-of-consciousness style, piling on impressions," Jill Neimark wrote in *Psychology Today*. She pointed to the results of a 1982 study that compared the responses from psychic sleuths, college students, and homicide detectives. "None of the three groups scored better than they would have if left to chance, but the psychics produced 10 times as much information, increasing their likelihood of a chance hit."

The authors of *The Blue Sense*, an exhaustive study of psychics and their work solving crime, concluded that the existing evidence of a "blue sense" did not yet meet the burden of proof, but they added that a lack of proof does not equal disproof, and that more study was required. While stories of psychics' abilities were exaggerated, they weren't as insubstantial as debunkers insisted, and the authors compared them in usefulness to FBI profilers. Interestingly, in a 2007 *New Yorker* piece about criminal profiling, the famed FBI profiler John Douglas is quoted as saying, "If there's a psychic component to this, I won't run from it." (Since this was a skeptical piece about profilers, Malcolm Gladwell, the author, likely intended that as further evidence that the work of profilers is suspect.) Today Dr. Arlan Andrews, a former White House Science Fellow who sits on the board of the current Rhine Research Center, says of many popular psychics, "At best, these are entertainers, not scientists. At worst they are frauds. Until the modern era, nobody has done any research into the causes and development of psychic talents, and even now there is no accepted theoretical basis for all the phenomena, nor proven methods of improvement." Although on that last point, the same could be said for ESP. *The Blue Sense* authors were unreserved in their opinion of Hurkos's claims in any case, which they summed up as "pure bunk."

Hurkos supporter Dr. Riesenman once said, "Give Peter Hurkos a scrap of clothing or a wisp of his hair and he might be able to tell you everything a man has done or is planning to do." And yet, when Hurkos went to Michigan to "work" on another series of unsolved murders, he gave a demonstration to a bunch of reporters in a restaurant while the actual murderer sat at a table nearby. Hurkos never sensed the murderer, who watched him from only a few feet away and dismissed him as "just a big clown."

Just because someone is a buffoon doesn't mean he isn't talented, and Hurkos may have had extraordinary abilities. But only in the most generous interpretation can it be said that Peter ever contributed anything significant to police investigations. Instead he traipsed around the country announcing he'd have a case solved in two weeks (it was always two weeks), leaving families more devastated and broken than they were before he appeared. He was either outright wrong, or pointed the police to suspects they were already considering. Hurkos's involvement never led to the arrest of the actual murderers. Even those who believe Hurkos is genuinely gifted must concede that his ability had no practical application in law enforcement. "If any use is made of such a man [a psychic sleuth]," Rhine summed up, "it would have to be considered as a sort of last-straw effort, and any findings he claims to present would have to be taken with a great deal of caution. Certainly no one should be charged or even put under suspicion by any statements made by a person like Hurkos."

A number of people, including *Chicago Tribune* reporter Norma Lee Browning, encouraged Rhine to test Hurkos, and Rhine was willing. But like Arthur Ford and every other medium except Eileen Garrett, Peter Hurkos never actually made it to the Parapsychology Lab.

A couple of months later, the Bridey Murphy hypnotist Morey Bernstein devised a rather lurid ESP test that he hoped would entice Hurkos to finally show up. Like most people, Bernstein thought the Zener cards were lifeless and dull and he couldn't imagine getting far with them with Hurkos. "I've observed that his mental radar detects most readily those events which are charged with emotion or action." So Bernstein had an artist make up an alternative set of ESP cards. "One card pictures an hysterical woman, bleeding from a stab wound—the knife protruding, of course. This one is in red. A second card has a vicious black panther springing directly at the viewer. And a third depicts a flaming airplane crashing to the earth. And so on." Although Peter and his associates said they were willing to give it a try, Bernstein had the same experience as Rhine—somehow a date was never made. "In my opinion," Morey wrote, "Hurkos will not submit to controlled tests by anyone who he knows will not tamper with the results."

Browning wrote in one of her books that Peter had tried many times to come to the Parapsychology Laboratory. "But for reasons known only to himself, Rhine has refused to accept Peter as a subject for his ESP experiments." Professor Rowe is blunt in his opinion of Browning. "I have always regarded her as Hurkos's sock puppet. She simply repeated what he told her and never did any fact-checking."

Hurkos did allow a Dr. Andrija Puharich to test him. Puharich was a doctor and a psychical researcher whose foundation was responsible for bringing Hurkos from the Netherlands to the United States in 1956. Rhine always felt uneasy about Andrija Puharich. While he did his best to maintain an open mind about others doing work in parapsychology, including Peter Hurkos, he was always openly hostile about Puharich.

Puharich had started out respectably enough. He was once invited by Charles Kettering (of Sloan-Kettering) to submit a proposal for research in "nervous phenomena." And in the fifties he placed Eileen Garrett in a Faraday cage and conducted experiments that yielded some interesting results. (A Faraday cage shields the interior against electrical static and electromagnetic radiation; for instance, a radio will not play inside a Faraday cage.) But at one point "he lost track of the serious research," Garrett's daughter Eileen and granddaughter Lisette say. "Puharich was directly connected to a number of psychic individuals, or at least some who claimed to be psychic, who were of somewhat bad repute," writes Joseph W. McMoneagle, a former army officer who had been recruited in 1977 into Stargate, the Army's remote-viewing program.

Puharich was also friends with Ira Einhorn, who is best known for murdering his girlfriend Holly Maddux in 1977 and then storing her body in his closet for a year. Einhorn brought Holly to Puharich's house in Ossining, New York, a year before killing her. When Holly first went missing Puharich was asked to call Einhorn for information about Holly's whereabouts. "Puharich agreed," according to the book *The Unicorn's Secret*, then he "called back saying everything was 'cool.'" Puharich later warned Einhorn that "intelligence agencies had been spreading a rumor" that he had killed Holly. Einhorn always claimed that he was innocent and that government agents were framing him for Holly's murder. He said he didn't know that Holly's body was in his closet until the police found it.

Still, Puharich's friendship with Ira Einhorn didn't prove he was a terrible scientist. In the end, Rhine may have simply had a problem with the fact that Puharich was the kind of guy who took people like Hurkos seriously.

Peter Hurkos would slowly lose the trust of many of his backers, although many still believed in his abilities and felt more pity than anger. The former president of Hurkos's foundation wrote Bernstein, "It distresses me very much to see one of the truly great psychics of our age get himself into such a mess." Rhine was asked if he'd be interested in an article disproving Hurkos's claims for the *Journal of Parapsychology*. But Rhine didn't want to play the role of the debunker. "I found that exposing these people did not put them out of business." Also, as he explained to his friend Morey, "From all I know, you think he has, or at least had, a lot of ESP ability. I'm inclined to that thought myself, but you must remember that I think everyone potentially has this ability, and being a charlatan at the same time need not rule one out." In 1964 Peter got into trouble with the California Parapsychology Foundation. "The man makes promises of cooperation in research, etc., but promptly breaks them," the foundation's founder wrote. "We are simply shocked by his lack of ethics where money is concerned." Although Peter had once predicted that he would die in 1961, he died in 1988, when he was seventy-seven.

Rhine would always have mixed feelings about mediums and psychics. It was possible that they had the abilities they claimed, he believed, but mediums were trouble. They were frequently alcoholic. They tended toward promiscuity. And if they were men they were often gay, and while Rhine was actually quite gay-friendly, he was still a product of his time and may have been a little thrown by this. Rhine once wrote Upton Sinclair, advising him to downplay the mediumship parts of his books, but Upton's wife, Mary Craig, objected strongly. "The cost to us of this unpopular matter is worth paying," she wrote. "Prestige has only one value: to use it to spread knowledge." But what knowledge, Rhine asserted. There simply wasn't a lot of hard data about psychics because they were impossible to get into the lab.

In spite of his attempts to keep his distance from psychics such as Peter Hurkos, like the father of Bruce Kremen, people continued to write Rhine as if he had a stable of people like Hurkos on his staff. When a jet crashed on Long Island in 1961, killing six, the assistant vice president of safety of American Airlines wrote a very careful and tentative letter asking if extrasensory perception might be of any use solving this and other unsolved aviation accidents. They had some personal effects of the crew members that had washed up on shore, along with their bodies, part of the cockpit, and one percent of the rest of the plane. "I would like to explore with you on a strictly confidential basis, the feasibility of attempting to find out what happened in the cockpit just prior to our Boeing 707 accident at Amagansett Beach, Long Island on January 26, 1961." Rhine told him the truth. "The stage at which research in parapsychology has arrived may be likened to a bridge which is still lacking only a single span for completion." Letters asking for help still arrived daily. A woman in California sent a telegram about her thirteen-year-old daughter who had run away. A girl in Wheeling, West Virginia, asked for help finding her friend who disappeared on her way to work. Rhine wrote back again and again with what would become his standard response. If people do have this ability there is no evidence that they can control it, and so the lab has given up the practice of giving out the names of psychics.

NINE

It was the sixties now, but the Parapsychology Laboratory was still sending out ESP cards while other parapsychologists had begun to branch out into newer and more alluring areas. Karlis Osis, for instance, who had left the lab to work for Eileen Garrett (she often picked up the lab's strays), came out with a report in 1961 called "Deathbed Observations by Physicians and Nurses." Osis had been collecting such accounts for the past two years, and what would come to be known as near-death experiences would go on to become a blockbuster area of research.

Osis had been inspired by a story in a 1924 book called *Deathbed Visions*. A dying woman had not been told that her sister, Vida, had died the month before. As the end neared, the woman told her mother that she saw her father, who had died, but then she looked confused and said, "He has Vida with him." Since as far as she knew, Vida was alive, it didn't make sense. She turned to her mother and said with surprise, "Vida is with him!" The woman died shortly after. The story was made all the more convincing to Osis because it had been reported by a hospital surgeon. Reasoning that their accounts would be more objective, Osis sent out ten thousand questionnaires about near-death experiences to medical professionals. He got six hundred forty back.

Just under forty percent of the people in the cases reported saw visions. However, they were not always the inviting images we're used to hearing about. In one account, "The patient had a horrified expression, turned his head in all directions and said, 'Hell, Hell, all I see is Hell.'" But some visions were so beautiful the patients were eager to die.

Rhine read Karlis Osis's report, but the Parapsychology Lab decided not to get involved in near-death experience research and continued to focus on ESP. As fascinating as the near-death accounts were, they were a return to the survival question that Rhine believed would stand in the way of science embracing their field. On the other hand, if they could crack open the mystery behind their intriguing results, if they could find a way to enhance and control telepathy and psychokinesis even a little, science would pay attention and things would get a lot more exciting.

Around this time Rhine received an invitation to participate in a Harvard Law School Forum titled ESP. The first three psychologists at Harvard who had been invited refused, but one young psychologist jumped on it—Timothy Leary. Leary was sharing an office with a Harvard professor named Richard Alpert at the time. The two men would soon make a name for themselves by experimenting with psilocybin, but there was a buzz about them from the start. "They had a cadre of acolytes even before they were introduced to the 'magic mushrooms,'" a former Harvard student remembered. A friend of Rhine's at Harvard described Leary as "tart and humorous," and encouraged Rhine to join him on the panel. It was to take place on April 14, 1961. Rhine accepted.

People had been after Rhine about drugs for a long time, beginning with John Collier, the commissioner of the Bureau of Indian Affairs, who sent Rhine a supply of peyote in 1936. Although they had studied the effects of various stimulants and depressants on ESP, Rhine was generally very cautious about experimenting with drugs. There were some indications, however, that he could be more open given the right circumstances. In 1954, after Huxley published *Doors of Perception*, the book that chronicled his mescaline experiences, Rhine wrote, "Perhaps we will follow you through this mescaline door with some psi tests." A few years later, Rhine broached the subject with Huxley again. Had he heard anything about LSD "releasing a subject's psi capacities," Rhine asked. Yes. Humphrey Osmond, who had supplied the mescaline for Huxley's earlier experiments, had "found that there seemed to be telepathic rapports between himself and another man, while they were both under the influence of the drug." But designing an experiment to be conducted with someone who has taken LSD was proving to be somewhat challenging. "It is rather like asking somebody who is listening with rapt attention to a Bach Prelude and Fugue, or is in the midst of making love, to answer a questionnaire."

Early in 1961, Timothy Leary sought out Aldous Huxley, who was then a visiting professor at MIT. Huxley told Leary about Rhine. Arthur Koestler, a friend of both Huxley's and Leary's, was visiting the Parapsychology Laboratory, and he told Rhine about Leary. By February, J. B. Rhine, who had never so much as taken a drink, was making plans to have Timothy Leary visit the lab after the Harvard panel, and to bring some psilocybin pills with him. Perhaps these fantastical drugs were the key to unlocking the mystery behind ESP.

At the ESP forum at Harvard, Leary had been more than amusing. He was fearless. In front of all the attendees, and all his peers and students, Leary got up and was unequivocally positive about the Parapsychology Lab and all their work. For Rhine the most important thing was he now had a colleague at Harvard who was interested in ESP. Timothy Leary just might be the one to help them move the research along. Leary was similarly electrified. "Big doings with Rhine," Leary wrote Arthur Koestler. "Overflow audience. Much excitement and very favorable reaction." Both men had already noted their differences in approach, however. "I spent a day with him," Leary continued in his letter to Koestler, "and tried to get him out of the old laboratory routine," but no such luck. Each was sure he could bring the other one over to his way of doing things. "I'm going to Duke next month to give them the mushroom," Leary told Koestler. Rhine, in turn, sent Leary ESP cards.

At the beginning of June, Timothy Leary and Dick Alpert flew down to Duke. On the morning of June 3, 1961, Rhine and a group of the lab research associates took the mushrooms. After spending what Leary called "six affectionate and sacred hours," he and Alpert were already on to their next appointment, leaving behind a form with questions that they asked everyone to fill out three days after they took the drug. The questions were:

1. Summarize in a few lines the most important or vivid characteristic aspect of your drug experience.

2. What in your life was the experience most like?

3. Discuss briefly your thoughts and feelings about the session in the two or three days following. (Also note if you did not think about it.)

4. Describe briefly how you would like to use the drug experience in future sessions.

Twenty-fours hours after what he called "the high point of the 'excursion,'" Rhine sat down and wrote Leary about their "boat ride." Overall the experience hadn't been terribly dramatic for him. He felt drowsy and dizzy and noticed "some awkwardness of movement and perceptual distortion such as the way faces looked under concentrated staring." And later he felt "a slight suggestion of the euphoria and stimulation that I recalled from dexedrine," but for the most part, he didn't have a strong physical response. The emotional response was more memorable. He called the feeling of togetherness that they all experienced "deep and genuine," and while he expected the feeling to pass, the next morning he felt a strong "wish to round up the boat crew and take off on some common endeavor, drug or no drug." What stood out for Rhine was that "we understood each other so readily. This is the cognitive side of the togetherness." That said, Rhine still wanted to see what, if anything, persisted from the experience.

What Rhine and Leary had understood so well, it turns out, was how different they were. Even while mildly hallucinating, the men debated. For Rhine the experiment was about the "larger interest of science and man's need to understand and control himself." But Alpert had warned that "we must not make the boat ride a structure of experimental design or the boat might as well be tied up permanently at the wharf." Rhine thought the Harvard psychologists were in danger of being permanently out to sea. The issue of lab experiments versus field investigations was an old and familiar conflict for Rhine and the central difference between himself and Leary and Alpert. "There is something about this that is symbolic of the whole program of the psychological sciences, the study of persons, human beings," Rhine wrote. "You have to get on the boat with them and away from the investigation pier to find out what they are like; but where would you ever get if it were all boat ride and never a pier with its equipment, design and controls?" Rhine also had a purpose and could not forget that "no matter how far out of sight of land we went, that this was a parapsychological tour, after all . . . and the question had to come out that has been living with us day and night, year in and year out: How do these odd capacities operate? How can we liberate them?" Tripping might be fun, but if they couldn't learn anything from the voyage, Rhine would just as soon stay home. However, Leary believed the mushrooms could liberate the "odd capacities." They "serve the single but important function," he wrote, "of 'unplugging' the mental-verbal-mind so that the ESP communication can occur."

Hallucinogens were fun and not yet illegal, and some of the staff members were eager to try the drug again. Rhine wrote Leary that he was repeatedly asked, "When are Dick and Tim coming again?" But not everyone at the lab was enamored of Leary. Rhine's daughter Sally said that Leary had "made some soupy romantic statement about the state of pregnancy that I remember really turned me off." But the drugs were unknown and thrilling, and even Xerox inventor and lab contributor Chester Carlson wrote to Rhine that a psychiatrist on the board of the American Society for Psychical Research wanted to give him LSD, and "Do you see any harm or danger in that?"

Rhine decided to conduct another psilocybin experiment. "On Saturday, the tenth, we used up the bottle of pills you left us." This time Gaither, Louie, and Bill Roll were members of the group. Louie was affected physically. She didn't feel sick exactly, but she didn't feel good either. John Altrocchi, one of the younger participants, described her. "Every once in a while she would raise her head (she was usually resting her head on her arms which were on the table) and look up rather disbelievingly at J.B. who was rather talkative throughout the whole thing. Then she would say something like, 'If that man took the same thing I took, I don't see how he can want to talk so much.'" Rhine asked everyone if they wanted him to read a paper on existentialism. "No," was the general response, although most failed to even respond. Rhine began anyway. You're not going to make it through one page, Altrocchi told him. Rhine made it as far as a paragraph. Gaither "went into it and came out it the thorough skeptic," Altrocchi said. He hadn't felt anything except some discomfort and a loss of some of his faculties. "There is no question but that he was much more uncomfortable than the rest of us throughout." Gaither also insisted that the seeming positive effects were "nothing but the results of suggestion and our imagination." The rest of the group thought "he was protesting far too much," but "nevertheless, one must admit that he may have a point."

Arthur Koestler, who had also taken psilocybin with Leary, sided with Gaither. "Chemically-induced raptures may be frightening or wonderfully gratifying; in either case they are in the nature of confidence tricks played on one's own nervous system."

Rhine seemed to agree with Gaither too. Even though he included a brief mention of their psilocybin experiments in his regular activity report to Duke President Dr. Daryl Hart, whenever people asked if he was experimenting with psilocybin or LSD he answered that there wasn't much there to encourage exploration.

Later Leary sent them a second questionnaire. It began with a brief description of his study, which would come to be known as the Harvard Psilocybin Project, and followed with a few questions that were more personal than last time:

■ How would you rate yourself on flexibility?

■ To what extent do you normally experience real joy in living?

■ Did you learn a lot about yourself and the world?

■ Has the mushroom experience changed you and your life?

Leary was still interested in conducting ESP research, but he envisioned a different sort of ESP test. "Imagine two people (be-mushroomed) in separate rooms communicating to each other, perhaps talking (into tape recorders) perhaps using colors or drawings," he wrote Rhine. "If we record the behavior of each person (as perceived by the other) it would become clear that a 'conversation' had taken place. . . . ESP communication," he went on, "will best be measured if the context is naturalistic (as opposed to experimental) and relevant to the deepest concerns of both (rather than scores on tests)." Which was what Eileen Garrett had always maintained.

But Rhine still had his eye on getting parapsychology admitted into the American Association for the Advancement of Science (AAAS), and for that they still needed experiments that could be more reliably replicated. Even if hallucinogens could enhance ESP abilities, they had to come up with ways to control the experiments in order for others to be able to precisely re-create them. Rhine continued to believe that Leary could help them further their aims. Leary had an extremely powerful personality, and that had always proved effective in evincing ESP. "You might find it amusing to play around with a simple Ouija board and have the group concentrate on influencing the board in some way," Rhine suggested to him. Then he outlined some simple controls, like using two rooms and outsiders to observe and signal the two groups. There is no record of Leary having ever performed this experiment.

On October 2, 1961, the lab applied to Sandoz Pharmaceuticals for psilocybin. Sandoz had developed the drug, and they were supplying it for free to legitimate researchers. But the only other lab reference to psilocybin tests was a letter Rhine wrote to Chester Carlson in December about a visiting doctor and associate professor in psychiatry who "did an experiment with psilocybin last Friday that has me worried, because he used five students." A lab research associate had been present as a supervisor, "but I am concerned because I was not sufficiently well-informed to have reminded all concerned of the necessary precautions." While Rhine did note that "the scoring rates went up," it looks like they never experimented with hallucinogens again.

Rhine may have made up his mind that psilocybin and LSD weren't worth pursuing, but for almost everyone else around him the experiments had barely begun. From all over the country people wrote, "Are you trying LSD?" Ivan Tors, a television producer friend of Rhine's who at the time was working on a new show called *Flipper*, wrote Rhine, "I, personally, experienced miraculous ESP faculties after the mushroom experience."

Eileen Garrett, who had been publishing pieces about psychedelics in *Tomorrow* since 1954 and had sponsored conferences on psychedelics and ESP in New York and France, went up to Harvard to meet with Timothy Leary herself. She had taken LSD on a number of occasions and wrote, "I believe that the drug has made me a better, more accurate sensitive when I perceive, hear, think and feel," although judging from her other writings, her experiences were decidedly mixed. Her granddaughter talked about the time Eileen tripped and people followed her with clipboards, recording her comments to a tree. Another time after taking LSD Eileen saw Aldous Huxley's wife, Maria, who had died in 1955. "I even felt her pulse, held her arm and found her living and breathing—her flesh warm," she wrote Huxley. "Though I have seen ghosts and held hands materialized . . . I could not on oath remember if they were warm or cold."

Eileen believed repeatable experiments using hallucinogens could be found. There was also the possibility that the drugs could help her learn more about her abilities. It had been twenty-five years since Rhine and Gaither attempted to study her, and she wasn't any closer to understanding the source of her visions or who Uvani or any of the other spirits who came to her were. Discarnate beings? Or was she truly crazy? She once quipped, "on Monday, Wednesday and Friday I think that they are actually what they claim to be . . . on Tuesday, Thursday and Saturday, I think they are multiple personality split-offs I have invented to make my work easier . . . on Sunday I try not to think about the problem." After meeting with Leary, Garrett gave him a grant and started storing LSD in the office safe. "She liked being associated professionally with a Harvard scientist," her granddaughter Lisette Coly said, "but she cut him off when he started all the tune in, drop out stuff. She didn't want any unscientific associations." Ultimately, she didn't get much further than Rhine.

The attitude in America toward hallucinogens was also beginning to change. By early 1962, when Rhine's friend Stanley Krippner went to Harvard to take psilocybin, Harvard and the local press had begun to oppose Leary's psilocybin project. The experiments had been postponed indefinitely. "Psychologists can be the most narrow-minded people in the world," Krippner wrote to Rhine. Rhine's answer said it all about his current feelings about the drug: "I note that you are thinking of another boat ride in the sea of fantasy. Is there a rudder and a compass?" Today Krippner agrees with Rhine's decision not to pursue experiments with psychedelics. "They're too unpredictable to serve a useful purpose in a controlled experiment," he says. There's no money for research in any case, he points out, because "there's no money to be made. No application."

By September 1962, when people asked Rhine what was new with Leary, he had to admit that he hadn't kept up with him. The last time Rhine had heard from him, Leary said that he had a talented group at work on extrasensory perception, but he didn't give Rhine any details. The next year, Leary's and Alpert's contracts at Harvard were not renewed. Rhine wrote to express sympathy. Their mutual friend, Aldous Huxley, died later that year, and after that Rhine and Leary fell completely out of touch. From then on, whenever asked, Rhine would deny working with psilocybin (or LSD), and while it's technically true, it's meaningful that he doesn't mention that he had taken it. Twice.

Rhine and Leary weren't the first researchers to take an interest in a possible connection between the hallucinogenic properties of mushrooms and parapsychology, and a number of people had begun to experiment around the same time. In 1961 Karlis Osis reported giving LSD to six mediums and conducting experiments in psychometry (the medium tries to get information about a person by holding an object that belonged to them). One medium did well, but the others were too distracted by the trip to focus. It looks like the United States military and Andrija Puharich may have begun clinical testing even earlier. Puharich had written Rhine back in 1953 claiming that he'd found a way to significantly increase ESP performance, and that he was working with the staff at the Army's Chemical Center in Maryland to prepare a demonstration for the CIA. Everything was top secret, Puharich said, and he couldn't give Rhine any specific details, but he couldn't resist adding that he'd had a "most enthusiastic" audience of high-ranking officers of the Psychological Warfare Departments of the Army, Air Force, Navy, and the Pentagon. Rhine tried to find out just what involvement Puharich had with the military, but military officials would always deny having anything to do with him. "Puharich is not employed by the U.S. Army in any capacity as a consultant, prime contractor, or subcontractor," a colonel from the Office of the Chief of Research and Development insisted years later. Further, the Army "is not now performing any investigations or letting any contracts for research in the field of extra-sensory perception, nor are there any plans to do so." But Puharich had, in fact, been a captain working out of the Army Chemical Center in Maryland at the time he wrote Rhine about his successful demonstration. Then in 1959 Puharich published a book called *The Sacred Mushroom*. A couple of years later, on January 24, 1961, the TV show *One Step Beyond* aired an episode about ESP and psychedelics. Across America people turned on their TV sets and watched a small group of subjects eat hallucinogenic mushrooms. Host John Newland later took mushrooms a second time and Puharich administered ESP tests. During the show it was mentioned that Puharich was working with the U.S. Army.

Not long after the show aired, Barbara B. Brown, a scientist at Riker Laboratories in California, wrote to ask that her name be taken off a paper that Puharich had submitted to the *Journal of Parapsychology*. Sensing a kindred hostile-to-Puharich spirit, Rhine immediately wrote back. He'd heard that Riker laboratories had sent an expedition to an area in Mexico where hallucinogenic mushrooms were known to grow, and that it was part of an Army contract that involved Andrija Puharich. Could she tell him anything about it? Brown (who had appeared on the *One Step Beyond* program with Puharich and who would become famous in the 1970s for her research on biofeedback) wrote back, "I have been advised by the Army authorities that their interests are not to be discussed under any circumstances, so I would appreciate it if you would forget any association between Army Chemical and me!"

Meanwhile, the lab and the U.S. military learned that a parapsychology laboratory had been established in one of the large Russian state universities and research in telepathy had been going on for years. Given what Puharich was promising he could accomplish with ESP, and the fact that the Russians had begun experimenting, it's not hard to imagine Puharich didn't at least get a hearing from the military, if not more.

In the end, if a practical application presented itself, the military was willing to give almost anything, even ESP and LSD, a try. The Army admitted to Rhine that they were "interested in chemicals which induce hallucinogenic properties." The Navy also confided that they had conducted pharmacological research and some tests with hypnosis. A desirable application had presented itself, and that was control.

According to author Jon Ronson in *The Men Who Stare at Goats*, Allen Dulles, the director of the CIA, addressed his Princeton alumni group in 1953, the same year that Puharich claimed to have met with the CIA. Dulles announced that "Mind warfare is the great battlefield of the Cold War, and we have to do whatever it takes to win this." "But mind to Allen Dulles and mind to J. B. Rhine were two completely different things," says former Army remote viewer Joseph W. McMoneagle. "Rhine was looking at mind from a paranormal standpoint. Dulles was looking at mind as something to conquer, to subjugate, win over in a contest between two political ways of thinking and acting. Dulles and his crew were into [overseeing experiments] using drugs and poisons, not that weird 'shit' called paranormal. If you ever brought the subject up with them, they'd throw you out a second story window. I know, because I used to work with them."

Under the psychiatrist Sidney Gottlieb, the CIA was already experimenting with hypnosis by 1954 to see how it could be used as a form of psychological control. A few years later *Newsweek* ran a piece which read, "Fantastic as it sounds, a serious psychological-research project being conducted for the Joint Chiefs of Staff is a study of the possible use of extra-sensory perception. Those in on it are looking into the possibilities of using ESP not only to read the minds of Soviet leaders but to influence their thinking by long-range thought control." The Pentagon issued an immediate denial. The next year, then Vice President Richard Nixon asked experts at the University of Michigan's Mental Health Research Institute to make a report about the behavioral sciences. The scientists' subsequent statement included a warning, "the probability of a breakthrough in the control of the attitudes and beliefs of human beings" might be developed as a weapon by the communists, using "exceptionally effective educational techniques, drugs, subliminal stimulation, manipulations of motives, or some as yet unrecognized medium." Unless the West did something about it first. The CIA began testing hallucinogenic drugs in the 1950s and 1960s under the project name MK-Ultra (also under Gottlieb), which was exposed in the 1970s and was shown to have led to at least one death.

But ESP was something else, and there aren't any records of extensive research until the seventies, although there were always pockets of interest in ESP among the military and the U.S. government—a scientist here, a lab there. In 1957 the CIA conducted what might have been the first remote viewing experiment. A hypnotized subject described a new Soviet ballistic missile in enough technical detail to convince the aircraft plant research chief and Washington experts listening to tapes later to say there was "just enough substantive data to stimulate the imagination." However, it was decided that clairvoyance was too risky for collecting enemy guided missile data, and the section ended with "The mystery remains unsolved." A similar test was conducted by the Air Force in 1960. A colonel wrote that "we have been doing some work here in 'sending' subjects to various geographical areas under the influence of hypnosis." The subjects were student officers from Maxwell Air Force Base. They were "sent" to see if they could come back with the tail numbers from parked aircraft. Rhine responded by suggesting a modified version of his card test, something in between his ESP cards and Upton Sinclair's wife's test with drawings, but the colonel couldn't get his superiors interested in further experiments.

Some military researchers were concerned about the ramifications of Rhine's assertions about parapsychology. "Basically, your comments on psychokinesis disturb me a great deal," the manager of the Space Guidance Staff of the Hughes Aircraft Company's Advanced Projects Laboratory wrote Rhine. They were studying the effectiveness of weapons systems, and if psychokinesis was real, someone could influence the outcome of their experiments (and presumably the actual weapons systems in the field).

Astoundingly, the same year that Puharich was experimenting with mushrooms and ESP, and the military was vehemently denying working with him, the Air Force built a machine for testing ESP. Rhine had always dreamed of creating an ESP machine, which would take the place of the human sender and make their experiments more immune to fraud and delusion. Rhine had even contacted IBM about it as far back as 1938. IBM was initially excited. "There is no question in my mind," one employee wrote back enthusiastically, that "it would be possible to develop a machine along the lines that you outline." IBM even went ahead and designed preliminary forms and sent them to Rhine for comment. They were going to take up the matter with their engineering department and get back to him, but nothing came of it. (With Rhine's help, IBM in Canada did eventually conduct an ESP test in 1961. A card was inserted in the July 29, 1961, issue of the weekly magazine *Maclean's* and readers were given instructions, but the results were inconclusive, and the experiment was not refined or repeated.) In 1948 Rhine had written the president of Duke about an ESP machine that he had seen at Bell Laboratories. Rhine hoped that Bell Labs might build one for him too, but nothing came of that either. In the fifties Rhine learned that Westinghouse had built a small machine for testing ESP. "We believe we are on the track of a method to strengthen the ESP effect," Rhine's contact at Westinghouse proclaimed. But while the Westinghouse representative was willing to talk to Rhine, he was being tight-lipped about the project publicly. Rhine thought it was out of fear of competition. But the contact wrote back and said he was keeping quiet "out of fear of ridicule."

The Air Force Cambridge Research Laboratories started tests of their ESP machine in 1961 and issued a report in 1963, stating that they wanted "to establish within the Air Force, and therefore within the Department of Defense, an authoritative knowledge and a base of experience in the field of parapsychology." An electronics engineer named Everett F. Dagle designed an ESP machine, a random number generating computer called VERITAC, and he and a small group of scientists ran eighty thousand trials over one year with forty-five college girls from Endicott Junior College in Beverly, Massachusetts.

The report began with an acknowledgment about what an emotional and controversial subject ESP is, and how unnecessary that was because "certainly, it offers no threat of any kind." But the Air Force's paper was "not intended to give aid or comfort to either faction in the ESP controversy, but, rather, it is a straightforward report of an ESP experiment, the design of which is offered as an acceptable model for work in the field." The report also said that objective and emotionally unbiased scientists were needed for the work. To Rhine the report essentially dismissed him and his colleagues and the model that had fathered parapsychology. Worse, no one in the Air Force's experiment scored above chance.

The lab issued a snarky and condescending write-up of the Air Force experiment in their *Parapsychology Bulletin*. "The authors of the report indicate an exalted appreciation of their experimental plan so that even though it failed to capture any evidence of ESP, it is highly regarded by them as a model apparatus for the purpose." Dagle felt he was being unfairly lumped in with critics who had written off ESP without giving it a try. But if parapsychology was going to be accepted by the AAAS, something Dagle knew was important to Rhine, he argued that "an experiment which produces negative results, even one such as ours, should not be considered a disaster or an attempt to refute the ESP hypothesis," because "if, on a second experiment, with improved equipment and refined techniques, I produce highly positive results, they will be more readily accepted than if I had produced them on the first time around." If the government didn't develop some expertise themselves, Dagle argued, they were never going to provide significant support to independent researchers. "If we don't understand it, we'll oppose it," he wrote, characterizing their attitude. Dagle's superiors apparently opposed ESP research—they declined to support any additional experiments.

Rhine had been trying for years to get the military to make an investment in the research, but he had a different application in mind than control. Defense, he wrote in 1957, "ideally should be a fluid, sensitive, vigilant, anxiety-laden concept," and "the best defense is prevention." Rhine believed that parapsychology could be the nation's best guard against a surprise attack. As intelligence gatherers, telepathic Americans could "view" into the minds of their enemies and see their plans. Rhine and others believed that this was possible. "Spies and informers of the future will operate with a degree of efficiency now scarcely imaginable," Aldous Huxley wrote in a 1955 *Esquire* article. That kind of development effort would require not only a huge investment but an equally huge leap of faith. The government had made that leap before, Rhine argued. "When the decision was made to risk billions on the great Manhattan Project," he wrote, "was it known for a certainty that a practical, usable atomic bomb could be made within a given time? Certainly not."

What the U.S. military needed right now, Rhine said, was the parapsychological "equivalent of a Sputnik." When the Russians launched the tiny satellite, the United States responded with an all-out effort and got a man on the moon. Rhine tried to convince his military contacts that the parapsychological Sputnik was already in orbit when news broke in 1961 of a parapsychology laboratory that had been established at the University of Leningrad. The Russian lab was only a couple of years old, but apparently the Russians had been investigating ESP for as long as the lab had at Duke (it turns out they'd been at it since at least 1916). For all we know, Rhine reasoned, the Russians could already have a practical military application. "If the Russian have a research program more developed than the one we heard about," he wrote to one Air Force official, it's not likely that they'd volunteer information about it.

A lot of individuals in the government "are of the private opinion that the potential payoff and possibilities of success are such as to justify an applied research program," the official responded. Publicly it was another story. No one in the military wanted to go on record advising parapsychological research.

Rhine wasn't going to wait for the government to find out what the Russians were up to. On May 21, 1962, the lab sent Gaither Pratt on a ten-day trip to Leningrad, among other places, to visit the Russian lab. He was received warmly by colleagues of Leonid Leonidovich Vasiliev, the director of the laboratory who joined them later. The Russian scientists immediately asked, "What is the attitude of scientists in the West toward research in this field?" Perhaps the Russians were having the same difficulties of acceptance and felt like academic outcasts too. They were familiar with the work at Duke and all nodded at the mention of Charlie Stuart's name. They'd read his paper evaluating the mediumistic material of Eileen Garrett. When it came to the issue of survival, however, the Russians sided with Rhine. Interest in survival, they told Gaither, would delay acceptance of psi, and they were very disdainful about the subject. They also thought if parapsychology had any chance, work would have to come from the hard sciences and not psychology. ESP research that was physiologically based, the Russians argued, would be more acceptable to science, and that was there they concentrated their efforts. While Upton Sinclair was calling ESP a mental radio (and before him, Mark Twain called it mental telegraphy), the Russians had been calling it "biological radio" since 1916.

The Russians had also already experimented with hallucinogens. Professor Terentjev told Pratt he had tried an ESP experiment with peyote. In his experiment, people who had taken peyote tried to identify objects concealed in boxes. One subject asked, "How did you manage to put such a large building into such a small box?" In the box was a stamp with a picture of the Telegraph Building in Moscow.

When Gaither returned, news of his trip was in all the papers, and Rhine decided to try once more to convince the government to take the Russians' work seriously. A few other scientists had made the attempt, and Rhine wasn't happy about who was representing their field. In a report to NASA, Dr. Eugene B. Konecci, the director of their Biotechnology and Human Research Office in Washington, wrote that the Russians "may be the first to put a human thought in orbit or achieve mind-to-mind communication with humans on the moon." But the report quoted, of all people, Andrija Puharich and Norma Lee Browning. "God help our space program," Rhine wrote Konecci when he read the report. "You could scarcely have scooped up a worse handful of trash."

Rhine wrote the Institute for Defense Analyses in Washington himself to tell them that the lab had firsthand evidence that the Russians were committed about parapsychology. "Someone ought to be assigned to examine the situation and keep it under observation," he advised. A correspondent of Rhine's was so excited by Rhine's efforts that he wrote him and said, "During World War II, Dr. Albert Einstein wrote a letter to President Roosevelt which brought about the establishment of the Manhattan Project. You, sir, are the Dr. Einstein of para-psychology."

Knowing the kind of money the government could throw at a project if they truly felt the need, Rhine may have envi-sioned being on the verge of developing the parapsychological equivalent of an atomic bomb.

But the U.S. government did not feel the kind of fear that paves the way for a Manhattan Project, and the Manhattan Project of parapsychology never happened. A *Chicago Daily News* reporter asked Rhine in 1962 if the defense industry had looked into applications of psi, and Rhine answered that there were a few projects here and there, "but evidently the time is not right for an open and sufficiently bold handling of the subject in this country." Gaither's trip and Rhine's efforts had not moved them.

The time wouldn't be right until the seventies, when the CIA heard again that the Russians had psychic spies, perhaps this time a little more convincingly, and they gave fifty thou-sand dollars to the research institute SRI International in 1975 to look into what would ultimately be called remote viewing. The CIA wanted to experiment very quietly and outside aca-demia, and this ultimately led to the Army's psychic spy unit called Stargate, established in 1978, and funded by Congress every year until it was terminated in 1995, over twenty mil-lion dollars later. (That may sound like a well-funded program, but Stargate's seventeen years of funding was the equivalent of a couple of days' worth of nuclear weapons research, develop-ment, testing, and production spending. In 1993 the psychol-ogist Sybo Schouten looked at a hundred years of investment for parapsychology and found it compared to two months of conventional psychological research.) According to Joe McMoneagle, one of the original remote viewers in the program, over the years the "CIA, DIA, DEA, NSA, FBI, NSC, Border Patrol, Secret Service, White House, and all of the intelligence services of the Department of Defense, including the Coast Guard," made use of Stargate's remote viewers.

But then Congress had the CIA take over the program in 1994. The CIA contracted the American Institutes for Research (AIR) to do a study analyzing Stargate's effectiveness. The American Institutes for Research brought in Professor Jessica Utts from the University of California and Professor Ray Hyman from the University of Oregon to conduct the statistical analysis, except the CIA didn't arrange for Utts and Hyman to get the necessary security clearances to access all of Stargate's operations. "They were given a couple boxes of files to review, which hardly represented twenty years of results," McMoneagle says, and "these files were hand-picked by the CIA." Utts and Hyman were told to make their report without talking to all of the remote viewers and without reviewing any previous studies other than a remote viewing study that had not included the Stargate Program. Utts found statistically significant results and evidence that the program had value as an intelligence collection tool; Hyman did not. Professor Utts wrote a rebuttal to Hyman's comments, but that was removed from the final report. If nothing else, the CIA couldn't have been thrilled about taking over a controversial, secret intelligence-gathering operation that was now known to everyone in the intelligence world. In 1984 McMoneagle received a Legion of Merit, the highest peacetime award given by the Army, for providing intelligence on one hundred sixty-nine separate occasions, "which could not be obtained from any other intelligence source," McMoneagle adds.

It's hard to imagine that the Army would hand out a Legion of Merit medal for information without value. In the end, the biggest problem must have been the lack of control. The information viewers like McMoneagle were collecting may have been valuable, but they couldn't master their ability to obtain it, nor was it always possible to interpret or verify what the viewers had seen. There was a reason Rhine preferred the controlled conditions of the laboratory, but for the military, remote viewing and ESP was only of interest—and use—in the field. Without this degree of control they may have felt it was pointless to pursue Stargate further. The potential must have been thrilling, though. How would an enemy defend themselves against remote viewing? There are no known shields against telepathy. The Army may have felt for a time like they were on the verge of finding the ultimate weapon of war.

TEN

The same year that Rhine was taking peyote with Timothy Leary, he got roped into attending a demonstration of an old-fashioned spiritualist standby: table rapping. The psychiatrist Dr. Kurt Fantl had been after him for five months to witness the work of a local "sensitive," and Rhine finally relented while on a visit to California in December 1961. It didn't go well. When Rhine inserted a few controls, little happened, and when it did, Rhine saw the sensitive's hand move. Rhine was gracious and nonaccusing and called the movement unconscious. However, he couldn't resist musing about how to improve the faked raps. "I think with a little rosin on the finger it would work better." (Rosin increases friction.)

Rhine had spent the last three decades moving parapsychology away from this sort of thing, but while he and the scientific community were happy to abandon the question of survival, the public's interest had not diminished. Scientists held firm to their beliefs about life after death, but the rest of the country was equally unwavering in theirs, and when it came to a battle of beliefs, if nothing else, science was outnumbered.

■

January 1964. Several groups of schoolchildren on field trips stand outside the Morris-Jumel Mansion in Washington Heights, New York, in the cold, waiting for a tour to begin. The oldest building in Manhattan, and once briefly the headquarters of George Washington, the mansion is now a historical landmark and museum managed by the Daughters of the American Revolution. It is named for the man who built it, Roger Morris, and the family who later made it famous, Eliza and Stephen Jumel. The tour is scheduled to begin at eleven a.m., and the children are waiting for the gardener who has the keys to open the door. Wendy Viernow, eight years old, is there with the group from Our Lady Queen of Martyrs, as are students from the local junior high school, Edward W. Stitt, also known as P.S. 164.

Just before eleven, a lady wearing a beautiful blue silk dress sweeps out onto the balcony. Her face is striking, and she looks down on the children with the air of someone who is used to commanding attention. All faces are tilted up when she speaks. The children are expecting a welcome, but instead the woman snaps at them. "Shut up!" she yells. Wendy remembers exchanging looks with the other kids. They were embarrassed and wounded. "We didn't think we were being bad," Wendy, who is fifty-one now, remembers of her field trip many years ago. When they look back up the lady is gone. According to another person who was present that day, she glided back into the house.

The gathered children would shortly learn that the woman who had scolded them hadn't been among the living for almost a century. But standing there below the balcony they would have sworn she was real. There was no reason for them to think otherwise. She wasn't transparent, as ghosts are traditionally described. She looked every bit as real as anyone else and they thought nothing of her appearance. As far as they were concerned, she was an angry tour guide dressed in a period costume who had yelled at them.

When Emma Campbell, the curator, arrives, they immediately tell her about the lady. There's no lady, she insists. No one lives in the house. We have just unlocked the door, she points out, and we're the first people inside the mansion today, as you can clearly see. When the children insist that a lady yelled at them from the balcony, they're shown that the balcony is locked and chained. This drains the indignation right out of them. The children then describe her blue dress that was covered with stars. Mrs. Campbell carefully explains that the dress is in a closet upstairs right now. It belonged to Eliza Jumel, the lady of the house. But she's been dead for ninety-nine years.

Almost from the moment the mansion was built, there have been reports of ghosts. Rumor has it that the stories were so well known that when Eliza and her husband, Stephen, bought the place in 1810, Eliza was able to get the owners to come down on the price. Anyone investigating the source of the haunting today has a lot of tormented dead people to choose from, including slaves, fallen soldiers, and a recurring theme of thwarted love. The Daughters of the American Revolution had a film that showed the outline of a female figure in period dress. They claimed it was the image of Sir Henry Clinton's mistress, who had six children by the British general before he deserted her after the revolution. There are a number of accounts of the ghost of a servant girl said to have killed herself by jumping from an upper-story window after having been cruelly used by a man of the house.

When the story of the apparition who appeared to the children hit the local papers, ghost hunter Hans Holzer immediately called Mrs. Campbell to offer his services. Holzer went to the mansion on January 19, 1964, with an entourage of a few journalists, members of the New-York Historical Society and the Daughters of the American Revolution, and his medium, Ethel Johnson Meyers. Johnson Meyers, a former opera singer, was a trance medium like Eileen Garrett, and her control was her dead husband, Albert. Whenever she went into a trance, Albert appeared, acting as the mediator between the dead and Hans Holzer, who regularly used Ethel's services. However, it was always a bittersweet and ultimately unsatisfying reunion for Ethel and Albert. Ethel was unconscious when Albert appeared, and when she awoke she remembered nothing. Hans Holzer spent more time with Albert than Ethel did.

Holzer and Johnson Meyers put on quite a performance for the journalists that day at the mansion. According to Holzer's account, Ethel walked into the Jumel house and immediately announced, "There is more than one disturbed person here. I almost feel as though three people were involved. There has been sickness and a change of heart. Someone got a raw deal." She then described a number of lingering spirits who, one by one overtook her and used her to speak. "I feel funny on my left side," she said as she walked around the mansion, limping. Ethel fell in front of a portrait of Eliza and complained in an imperious voice that her name should be on the picture. Mrs. Campbell whispered to Holzer that Eliza's name did once appear on a nameplate, but the nameplate had since been removed. Albert broke in at one point to interpret and explain what was happening on the other, unseen side. He indicated the portrait of Eliza and said that she had a guilty conscience.

At this point Emma Campbell took Holzer aside and repeated one of the most famous rumors about Eliza Jumel. "Stephen Jumel bled to death from a wound he had gotten in a carriage accident," she began. "Mme. Jumel allegedly tore off his bandage and let him die. That much we know." Eliza had murdered her husband, the reasoning went, in order to be able to marry Aaron Burr, the former vice president of the United States. It was a rumor that had been circulating for over one hundred thirty years, but Albert added a new detail. Stephen Jumel had been buried alive.

Life did not begin auspiciously for Eliza Jumel. She was born in a brothel in Providence, Rhode Island, in 1775. Her mother, Phebe, who sold the then five-year-old Eliza at a pauper's auction, was later murdered. Her sister Polly died before her twenty-fourth birthday. And while it's said that the famously beautiful Eliza had to prostitute herself in her youth, the various men who were taken with her beauty also provided her with a rudimentary education.

Eliza grew up and left Providence, traveling to Boston and France, where she continued to educate herself. Finally, she moved to New York and eventually married the French wine importer and merchant Stephen Jumel, who did everything he could to help elevate her, including buying her the mansion on the hill. Years later, when Stephen Jumel died, Eliza Jumel did in fact marry Aaron Burr. Eliza saved herself, escaping the fate of her mother, a woman so lost she once told the Providence town elders before they banished her, "I don't know where I was born."

At the end of her life, however, Eliza suffered from some sort of dementia and rarely left the mansion. Another story about her that was often repeated illustrates the distressing descent of her madness. It concerns a dinner table that many felt was the inspiration for a famous bleak character from a Charles Dickens novel.

Eliza had set a table for a fabulous dinner party; however, according to her once adored grandniece Eliza Jumel Chase, this particular dinner never happened—it was a fantasy of Eliza's. In her troubled mind, the dinner took place and it was a great success. Napoleon's older brother Joseph had been there, and everything about her dinner—the trays of sweets, the gold and silver ornaments—so impressed Bonaparte that she kept the table exactly as they left it. It would sit in the mansion uncleared for almost thirty years, decaying exactly like Miss Havisham's table in *Great Expectations*. After a time the odor faded, the food dried and mummified, and dust covered the table like a fog, giving the tableau a romantic soft focus if you didn't look too closely. It was a legend in the neighborhood.

Over the years the increasingly delusional Eliza became convinced that her family was trying to kill her, and she banished the last two family members who hadn't already either died or fled from her madness. She lived out her last years alone, her only regular human contact the few servants left in her employ. One October afternoon in 1862, a young neighbor named Anna Parker came to call with a few of her friends. After the visit, Parker went home and immediately wrote down everything that had happened in a letter. "There she stood on the front doorsteps . . . a more fearful looking old woman one seldom sees—her hair and teeth were false—her skin thick, and possessing no shadow of ever having been clear and handsome—her feet were enormous, and stockings, soiled and coarse, were in wrinkles over her shoes—on one foot she wore a gaiter and on the other she wore a carpet slipper. Her dress, or the skirt, which was all that was visible, was a dyed black silk with stamped flounces, three of them, such as were worn six or eight years ago. It was very rusty and narrow in the skirt. She wore a small hoop, which in sitting down she could not manage, so that it stood up, displaying her terrible feet. Over her shoulders she wore a rusty, threadbare black velvet talma—and a soiled white merino scarf around her neck—her cap was made of humbug white blonde and cotton black lace and had long pea-green streamers. Her appearance was anything but neat."

At one point Miss Parker and her companions passed by the remnants of Eliza's breakfast on the way to the parlor. "We were very much afraid that she would invite us to eat something," Anna wrote. Then they saw the famous decrepit dinner table they'd been hearing about since they were children: "on the left was the table—china, glass, still there, and gold ornaments and pyramids of confections, still standing on this greasy, dusty table, crumbled and moulded. This same table Mrs. Appleton Haven saw twenty years ago. It is unchanged now, except Madame was persuaded by Mrs. O'Conor that it was imprudent to leave so many gold and silver ornaments about, so some of them were put into the safe."

Soon after, Eliza closed the blinds and locked the mansion doors. In the summer of 1865, three years after Anna Parker's visit, Eliza would be dead, and all her family who had been banished from the house would return.

The ghost first appeared in the Washington Room. This had been Eliza's bedroom, and was now used by her grandniece Mrs. Eliza Jumel Pery and her husband, Paul. The Perys had a daughter, and her governess described the events during that period. "After a little time I was moved down to the Lafayette Room, to be nearer Mrs. Pery, who was in nightly terror of the ghost of Mrs. Jumel, which she claimed came with terrible rappings between twelve and one o'clock or about midnight." One night they roused a few servants to act as witnesses, then they all gathered to wait. "Suddenly there were loud raps like the sound of a mallet striking under the floor, and directly, seemingly, under

Mr. Pery's chair, from which he leapt as if he had been shot."
The sounds moved around the room. What began under Pery's
chair was "followed by a clatter of what sounded like a skeleton
hand drumming on the panes of the east front window." They
turned their heads as the drumming moved around the room
and then out into the hall. No one moved until they heard the
rappings come from the room where the Pery's daughter Mathilde was sleeping. And even then, it was the governess who
sprang to action and not Eliza Jumel Pery, Mathilde's mother.
"I stepped to the door and looked in," she described. "Even as
I looked, the tapping continued on the tin slop-pail and then
ceased altogether. The child was sleeping soundly and Mrs. Pery
thought I was very brave to enter the room at all."

Before she died Eliza changed her will and left most of her
money to her favorite charities, including the Association for
the Relief of Respectable, Aged, Indigent Females in the City
of New York; the New York Asylum for Orphans; the Society
for the Relief of the Destitute Children of Seamen; and the
Trustees of the Fund for Aged and Infirm Clergymen, among
others. The family successfully fought the will, although it
would take more than ten years. Nelson Chase, the husband of
Eliza's adopted daughter Mary, sold the mansion in 1887 and
committed suicide the following year.

A few months after Holzer's first visit to the Morris-Jumel
Mansion, Mrs. Campbell, the curator, called and asked if he'd
like to return on May 22, the anniversary of Stephen Jumel's
death. "I have always felt that anniversaries are good times to
solve murder cases," Holzer wrote in his account of the day.
He accepted her invitation. This time a reporter from the *New
York Times* was there, along with other print journalists; movie,
TV, and radio people; and members of the New-York Historical Society. In all, thirty people filled the room. They sat on

folding chairs facing the bed that Eliza Jumel had died on after asking a servant to make her up one last time.

Holzer began with a lecture explaining that ghosts were surviving emotional memories of human beings who died under traumatic circumstances. They were still occupying the physical side without realizing they weren't physical anymore. Poltergeists, for instance, who rapped or threw things, were actually panicked ghosts who were frightened because they were no longer noticed or seen. Like a therapist for the dead, Holzer had the job of calming them down and helping them cross over.

"Who are you?" Holzer asked, after Ethel Johnson Meyers had settled into her trance. "Je suis Stephen," Ethel/Stephen answered, while holding her body as if in pain. After that, the ghosts never answered a simple question that simply again.

"I'm alive," Ethel/Stephen insisted repeatedly throughout. "She took it, she took it—that woman. She took my life. Go away," Ethel/Stephen demanded, along with more talk about being buried alive. Then Ethel/Stephen uttered the name Aaron. That got Holzer's attention. He tried to get more information. "Aaron? Was he involved in your death?" Ethel/Stephen answered, "That strumpet . . . hold him! They buried me alive, I tell you." Then, "I'm bleeding." Holzer questioned him. "How did this happen?" "Pitchfork . . . wagon . . . hay . . ." "Was it an accident, yes or no?" "I fell on it." The only other thing Holzer could get out of him about the murder was the fact that a boy in the hay had pushed him off. There was a lot of talk about how "she wanted to be a lady," presumably talking about Eliza, the girl who had been born in a brothel. At one point Stephen called out "Mary, Mary!" (The name of their adopted daughter.) Holzer told him to go with Mary. "You have been revenged many times. She [Eliza] died miserably. Let go of this house."

Then a different voice took over. According to one of those present, the Ethel/Stephen voice had been "whining and complaining, like an old man in pain," but the next voice that took over and announced, "I am the wife of the Vice President of the United States. Leave my house!" was "arrogant and imperious." Holzer had an even harder time getting a straight answer out of Ethel/Eliza. "I will help you," he told her, "if you tell me what you did. Did you cause his death [Stephen's]?" "The rats that crawl," Ethel/Eliza answered, "they bite me. Where am I?" The conversation went nowhere. She repeated that she was the wife of the vice president of the United States, and called out names that no one knew. But like Ethel/Stephen, she also called out for Mary. Albert, the control, once again broke in to explain. "She's no longer in her right mind." He also revealed that she was guilty of Stephen's murder. Eliza was the one who had arranged for the boy to push Stephen off the wagon, and Aaron Burr was involved. "Burr believed Madame Jumel had more finances than she actually had." Albert said that Mary had led Stephen to the other side, but "Madame was not in a correct mental state to be exorcised." Holzer then asked Albert to bring Ethel back to her body.

Later Holzer tried to investigate whether or not Stephen Jumel had been buried alive. He spoke to two doctors and asked if there could be anything left of his body after having already been in the ground for one hundred thirty-two years. One doctor said yes and the other said no. Holzer approached the Manhattan District Attorney about exhuming Stephen Jumel. The DA told him to contact the Office of the Chief Medical Examiner, but the chief medical examiner at the time, Dr. Milton Halpern, turned him down. Holzer went to Old St. Patrick's Church at the corner of Prince and Mott streets, where Stephen Jumel is buried, but they wanted nothing to do with the business either.

Eliza Jumel, who suffered so at the end of her life and, according to Holzer's medium, continued to suffer in death, could have used a little compassion. But when it came to Eliza Jumel, Holzer was not a gentle crossover guide of the dead. Throughout the séance, he treated her with pitiless contempt. He ends his account of what happened with this parting shot: "Whether the Jumels, the remorseful Betsy [another name Eliza was known by] and the victimized Stephen, have since made up on the other side, is a moot question, and I doubt that Aaron Burr will want anything further to do with the, ah, lady, either." To this day you can hear the quotation marks whenever Holzer refers to Eliza as "the lady." Perhaps the fact that she may have been forced to become a prostitute as a child in order to survive offended his sensibilities. Holzer would probably say his attitude is perfectly appropriate for a murderess, but the truth is that Stephen Jumel died of more natural causes. He didn't bleed to death. His death certificate says that he died of inflammation of the lung (aka pneumonia).

Today, the Morris-Jumel Mansion is owned by the city of New York and managed by the nonprofit corporation Morris-Jumel, Inc. No one who works there now has ever experienced anything supernatural, but a biographer who wrote about Eliza Jumel in the 1970s said she sensed Eliza in the attic, and according to Patricia McMaster, who gives regular tours of the mansion, the weekend manager "said that male visitors have told her that they heard a woman's voice emanating from the clock in the first floor hallway instructing them to either 'come closer' or 'get out,' presumably depending on the attractiveness of the individual man!"

The more science explains the unexplainable, the greater our willingness to believe in the fantastic. Periodic Gallup polls throughout the years show consistently that our belief in the supernatural is increasing. In a 1978 Gallup poll, just under eleven percent of the respondents believed in ghosts. But their most recent poll (2005) found that belief has gone up to one person in three. Even more people believe in the devil (forty-two percent). On the Web site for the Committee for the Scientific Investigation of Claims of the Paranormal (CSICOP), Paul Kurtz, the chairman, calls the findings "disturbing." "They're polling the heart, not the head," a CSICOP senior research fellow added.

People love a good ghost story. They always have. Ghost stories and explanations of ghosts have been around for a long time. The Roman senator Pliny the Younger warned a friend against renting a certain house because "at the dead of night horrid noises were heard in the villa: the clashing of chains, which grew louder and louder." The next part reads like the ghost of Jacob Marley from *A Christmas Carol*. "Suddenly, the hideous phantom of an old man appeared who seemed the very picture of abject filth and misery. His beard was long and matted, his white hair disheveled. His thick legs were loaded down by heavy irons that he dragged wearily along with a painful moaning; his wrists were shackled by long, cruel links." Then he'd raise his arms and shake his chains, in "impotent fury." Karlis Osis pointed out that we don't see ghosts in chains so much anymore because we've stopping putting people in chains (mostly). An even earlier description of the sad lot of the undead is found in Plato's *Phaedo*, where he explains that apparitions are souls who are "polluted" and afraid and still in love with and attached to the body. His description of such souls as so depressed and "afraid of the invisible and of the world below" that they would prefer to prowl "about tombs and sepulchres" is, like so many ghost stories, more pitiable than scary.

Where there are believers, there have always been skeptics. In 1920 a New Yorker named Joseph F. Rinn offered five thousand dollars to the physicist and psychical researcher Sir Oliver Lodge for tangible evidence of communication with the beyond. Lodge said it was only natural that people were trying hard to break down the barrier between life and death, "since so many people have lost those closest to them through the ravages of war." He was referring to World War I, which had taken the lives of over eight million men worldwide. It was a motivation that, unfortunately, Sir Oliver was intimately acquainted with. He'd lost his youngest son, Raymond, to the war in 1915. In response to Rinn's challenge, Lodge said, perhaps with some regret, "You cannot get things by simply offering money."

The same year, 1964, that the Morris-Jumel Mansion haunting story hit the papers, magician and skeptic James Randi wrote J. B. Rhine and invited him to be a guest on a radio show about parapsychology, but Rhine declined. He had mixed feelings about the man who called himself the Amazing Randi. It was entertainment and not science. "A certain amount of clash of egos and contention between partisans is tolerable if there is in the background an evident interest in finding a solution that is right," Rhine wrote back. But "merely putting on a program because it is lively and without caring how it turns out would, I think, rob the program of interest for me and many, many others."

The "many others" Rhine mentions could be anyone. "Belief in these phenomena is not limited to a quirky handful on the lunatic fringe. It is more pervasive than most of us like to think, and this is curious considering how far science has come since the Middle Ages," Michael Shermer writes in his book *Why People Believe Weird Things*. In the end, perhaps not so curious. What could be more irresistible and comforting than the idea that death is not the end, and that if we didn't find what we were looking for this time around, we will always have another chance?

Still, parapsychologists like Bill Roll and his colleague William Joines try to bring parapsychological research forward. Like many other parapsychologists, Roll and Joines look at ways in which quantum theory might explain the paranormal, especially cases that have been open and unexplained for years. There was one case from 1961 in Newark, New Jersey, that Roll continues to study to this day. Mabelle Clark was living in a housing project with her grandson Ernest Rivers. Five years earlier, on December 14, 1956, Ernest's mother had murdered his father, a former Golden Gloves champion, in his sleep. Ernest was then eight years old. At four in the morning, Anne Rivers shook her husband and asked if he still didn't care about her, and when he didn't respond she shot him once in the back and once in his left side. She told the police that he had committed suicide, but confessed to the killing in a matter of hours. She said she did it because her husband kept telling her things like, "You are nothing but a doctor's bill to me." She went to prison and Mrs. Clark took over raising her grandson.

The disturbances began on May 6, Ernest's thirteenth birthday. His mother had escaped from prison and was expected to show up for his party. She did not. But that night bottles, cups, pepper shakers, ashtrays, and lightbulbs fell, flew, and moved about when there was no one there to move them. The newspapers got hold of the story almost immediately, and an amateur exorcist named Edward J. Del Russo came out to the apartment. He burned a candle beside a glass of water and announced to all assembled that the ghost was gone. "It's done. Nothing is going to happen. Everything is under control."

The disturbances resumed the next day. Mrs. Lucille Herrmann, the housewife from Seaford, Long Island, was called for a quote. She told Mrs. Clark to "keep up your courage and don't panic." Ernest's grandmother was afraid that the disturbances were caused by her dead son-in-law, who she said was a violent man. She thought he might be trying to get Ernest away from her.

Dr. Charles D. Wrege, an assistant professor in the Department of Management at Rutgers University who had an interest in parapsychology, investigated first. One night when he was alone with Ernest a bunch of drunks pounded on the front door. We want to see the flying ashtrays, they yelled. When Dr. Wrege wouldn't open the door, they threw a rock through the window. Wrege grabbed Ernest and ran into the kitchen. He kept one arm around the boy to comfort and protect him while he called the police, but Ernest remained extremely upset. Fifteen feet away, a lamp fell. Wrege determined that the disturbances could not be explained by known causes or fraud. The strange activity lasted for a little over a week, and ended when Ernest was sent to live with another relative.

Bill Roll got involved that September. He set up appointments for Ernest with psychologists at New York University. A few months later, the supervisor of the New Jersey Board of Child Welfare arranged for Ernest to be brought to Duke University for study, but by that time Roll already suspected that Ernest may have been causing some of the events in ordinary ways.

At the lab, Roll and Gaither Pratt put Ernest and his grandmother in a room with a one-way mirror and watched from the other side. Gaither saw Ernest hide two measuring tapes under his shirt and then throw them at his grandmother when she wasn't looking. Ernest denied throwing the tapes. He continued to deny it while under hypnosis, and again during a polygraph examination. The polygraph indicated he was telling the truth. One of the Duke psychologists, who had found indications of an "intense underlying anger toward his grandmother which he was unable to verbalize," suggested that Ernest had thrown the tapes while in a dissociated state. Roll concluded that the initial events were genuine, but the later events were fraudulent.

Part of the explanation for the initial events, and others like them, may have something to do with what Roll and Joines call psi entanglement. In modern physics, quantum entanglement occurs when action performed on one of two particles separated in space appears to instantaneously influence the other (also referred to as nonlocality). This has been proven experimentally, and leads Roll and Joines to consider that "something unaccounted for is connecting otherwise isolated objects." A lot of people turn to quantum entanglement when looking for an explanation for paranormal events. Even Bruce Rosenblum, a physics professor at the University of California at Santa Cruz, wrote in his book *Quantum Enigma*, "As yet, evidence for the existence of paraphenomena strong enough to convince skeptics does not exist. But if—*if!*—such a phenomenon were convincingly demonstrated, we know where to start looking for an explanation: the quantum effect of consciousness, Einstein's 'spooky interactions.'" Einstein had famously described quantum entanglement as "spooky action at a distance."

The idea of instantaneous action at a distance had no place in Einstein's understanding of the physical world. The speed of light is constant and events in one place cannot simultaneously influence events in another. "No reasonable definition of reality could be expected to permit this," he and two colleagues wrote in a 1935 paper. But French physicist Alain Aspect and others have since proved otherwise. It appears that separate particles can influence each other.

Einstein would have had the same problem with telepathy that he had with quantum entanglement and instant action at a distance. The experiments Gaither Pratt and Hubert Pearce conducted seemed to indicate that information traveling between two minds was also instantaneous, regardless of distance. Hubert scored as well whether he and Gaither were in the same room or in different buildings altogether. The philosopher Curt John Ducasse would later point out that perhaps Rhine and others were mistaken in their interpretation of Hubert and Gaither's experiments. After all, the timing wasn't really as precise as possible: it was two men checking their watches. And the distance wasn't all that great: Hubert and Gaither were separated by nothing more than a college quadrangle. To say that the sending and receiving was instantaneous was perhaps not correct. Ducasse also pointed out that Rhine could not measure the magnitude of the telepathic signal. Perhaps over some distance it had been reduced to a "telepathic whisper." That would mean that telepathy was subject to the laws of time and space, and this might have been acceptable to Einstein, had he been willing to accept ESP in the first place.

Rhine wrote Einstein in 1940 about "the problem of finding an adequate physical theory" to account for the results he was getting. But Einstein replied that while he had read his books, "I must confess I am very sceptical about the reality of the phenomena in question although I have no explanation for the positive results you have obtained together with your collaborators. In any case I don't feel able to contribute effectively to the elucidation of the problems concerned." But as noted earlier, six years later Einstein wrote parapsychologist Jan Ehrenwald that his book about telepathy "has been very stimulating for me, and it has somewhat 'softened' my originally quite negative attitude toward the whole of this complex of questions. One should not walk through the world wearing blinders." Three years after that, when another physicist said he was inclined to believe in telepathy, Einstein theorized that "this has probably more to do with physics than with psychology."

A telepathic whisper is no longer required to make telepathy acceptable to current laws of science. As spooky as it is, Alain Aspect and others have demonstrated the phenomenon known as nonlocality is real. "It may mean that everything in the universe is in kind of a rapport," physicist David Bohm once commented. Psi phenomena may simply be manifestations of this rapport. Princeton researchers Robert G. Jahn and Brenda J. Dunne have conducted remote perception experiments that separated the subjects by thousands of miles, and they didn't find any evidence of a time delay, indicating that remote perception, telepathy, consciousness—whatever you want to call it—may be nonlocal as well.

Science has made another unsettling discovery that Roll and Joines believe may one day help explain PK. Experiments have shown not only that one cannot observe both an electron's motion and position at the same time, but that the observer can affect what the electron becomes: a particle or a wave. The Nobel Laureate physicist Steven Weinberg wrote, "It's an unpleasant thing to bring people into the laws of physics." But the observer affects the outcome. Science believes that the measuring device is what affects the outcome, but others have suggested that it can only be the consciousness of the observer. The mathematician John von Neumann pointed out that everything in the physical world has to obey the laws of quantum physics, so therefore something nonphysical, like consciousness, has to explain how the observer affects the wave/particle outcome. Roll and Joines believe this opens the door to understanding PK. "The more we know about the movement of material objects without tangible aid," they wrote, "the more normal it seems, and the more we know about matter, the more paranormal it seems." While the physicist and skeptic Victor J. Stenger would call this "quantum quackery," as do most mainstream physicists, quantum theorist Euan Squires writes, "That there are 'connections' of some sort between conscious minds and physical matter is surely implied by the fact that conscious decisions have effects on matter."

There is perhaps another explanation. In the 1950s the physicist Hugh Everett proposed that the observer was not affecting whether or not the electron became a particle or a wave, but instead it had become both. He theorized that there were many universes, which were endlessly branching off, and in one universe the electron became a particle and in another it became a wave. If information can leak or be transmitted from one universe to another, Everett's theory may explain Rhine's results as well. "We are not big fans of the parallel worlds paradigm," Robert Jahn says, however. To explain what he calls "remote perception" Jahn prefers a model "derived from our 'Sensors, Filters, and the Source of Reality' theory, i.e., the percipients are utilizing different filters in their penetration into the ontological reality."

A common response to all of this is to limit the various phenomena to the quantum world, that is, not use this data to infer things about psi or anything else that happens in the macroscopic or visible world. But where is the line drawn? It happens here at this size, but not at this? Can this be pinpointed and demonstrated experimentally? The biologist Robert Lanza argues drawing a line in the sand like this "has no basis in reason, and is being challenged in laboratories around the world." Experiments have shown that entanglement, for instance, extends to the macroscopic world as well. Without knowing more about psi, it's hard to know where along the line it falls, or if it belongs on the line at all. Rhine, who didn't believe psi was physical, would argue that it doesn't. "When Dr. Rhine said that the phenomena of parapsychology did not follow the laws of physics, it was classical [Newtonian] physics he meant," Betty McMahan writes. "Quantum mechanics was scarcely known by anyone outside theoretical physics in those days. But now, every popular scientific magazine gives accounts of the totally counterintuitive phenomena studied by theoretical physicists. Quantum physics might appear to require that parapsychological phenomena exist."

Roll made another interesting observation, linking Garrett's idea about the emotional component of psi with another theory from Robert Jahn and Brenda Dunne. Roll, like Garrett and many others, noticed that there was an emotional element in poltergeist cases, and going back to his Newark case he found it to be present there as well. Ernest's mother murdered his father and he was living with a grandmother toward whom he felt hostile. Jahn and Dunne had introduced an idea called consciousness waves, which Roll and Joines feel can be applied to poltergeists. Things physically or emotionally closer to us are more important, generally, than things that are physically or emotionally far away, and when we focus our consciousness on an object close to us it may become charged, something along the lines of an electromagnetic charge. In Ernest's case, objects in his grandmother's home may have become charged with the tremendous feelings he didn't want, couldn't control, and tried to repress. Jahn believes that information between object (ashtray) and agent (Ernest) may be carried by another field, not currently accounted for by science, what he calls an "information field."

"We have to get consciousness and information into the

physical theories," Jahn sums up. "The process of remote perception has propagation and communication," he explains. Therefore, "the rules of science have to be expanded to include another field," he says, "an information field with consciousness as an information processor." The quantum physicist Henry P. Stapp wrote that the "new physics presents prima facie evidence that our human thoughts are linked to nature by nonlocal connections: what a person chooses to do in one region seems immediately to affect what is true elsewhere in the universe. This nonlocal aspect can be understood by conceiving the universe to be not a collection of tiny bits of matter, but rather a growing compendium of 'bits of information.'" Almost thirty years of research and thinking have led Jahn to what he believes is the only possible next step and that step is contained in the title of a paper he recently published with Brenda Dunne: *Change the Rules!* What they call consciousness-correlated physical phenomena have been documented repeatedly all over the world. But "their appearance and behavior display substantial departures from conventional scientific criteria." They replicate irregularly, and subjective parameters such as emotion, intention, and unconscious information processing appear to be part of it. The physicist Freeman Dyson said he believed telepathy to be real, but that science could never prove it. Jahn and Dunne instead say that if science is to understand these phenomena, it has to "expand its current paradigm to acknowledge and codify a proactive role for the mind in the establishment of physical events, and to accommodate the spectrum of empirically indicated subject correlates." "Anomalous phenomena are anomalous precisely because they don't seem to obey these 'laws,'" Dunne explains, referring to the laws that govern the physical world, like space and time, but instead "indicate that consciousness is capable at times of interacting directly with the physical environment without the imposition of these filters." Jahn and Dunne call this new direction a "Science of the Subjective."

In general, few parapsychologists investigate hauntings today, but the parapsychologist Lloyd Auerbach, who has a master's degree in parapsychology from JFK University, has been researching hauntings since 1979. In 1989 he founded the Office of Paranormal Investigations. According to Auerbach there are two kinds of hauntings, one that involves an intelligent, conscious being who can interact with the living, and the other that is like the audio artifacts EVP researcher Tom Butler calls "echoes of the past." Auerbach calls the optical equivalent "place memories." These apparitions are visual memories that have been imprinted on a location, like a "loop of video or audio tape playing itself over and over."

There's little funding for apparition research currently, and the investigations Auerbach undertakes, about fifteen to twenty-five a year, are usually at the request of a home or business owner who is experiencing an unpleasant haunting and wants it to stop. Like Holzer, Auerbach works with a medium, and between the two of them, they determine who the dead person is and why they are still here, and then they come up with a way to resolve the situation to the satisfaction of both the living and the dead.

Although Auerbach believes parapsychology is "primarily a social science," he does use some tools. He sometimes brings a device to detect temperature changes, but while cold spots are part of ghost folklore, he rarely comes across them. He also has an ion detector, but he doesn't usually find a correlation between air ionization and ghosts. The changes to the environment that seem most consistently connected to ghost activity are fluctuations in the electromagnetic field, which

Auerbach reads using a TriField Meter. The TriField Meter measures AC magnetic and electric fields, and RF/microwave, and was designed to detect fields that needed to be shielded in order to protect equipment operation or health (ie, excessive microwave exposure.) Using a TriField Meter, Auerbach once followed what looked like a moving electromagnetic field throughout a house. "It's possible that a ghost moving through the house could affect the electromagnetic fields," he says, "leaving behind an electromagnetic wake. . . . But you can't rely on technology," he warned. "Just because you have a change doesn't mean you have something paranormal happening." On one occasion the readings on the meter surged, but it turned out to be a nearby coffee maker that had just turned on. "You have to correlate the change to something."

Dr. Michael Persinger, a cognitive neuroscientist from Laurentian University in Canada, also noticed that areas associated with hauntings tended to be "electromagnetically noisy." He developed a helmet that generates very weak magnetic fields similar in intensity to those measured within haunted areas. When people put the helmet on in experiments they see things.

"The critical principle here is that the temporal pattern of the field generates the experience." Persinger and his associate researcher Don Hill have been gathering data and studying the magnetic fields at sites where people have had paranormal experiences like apparitions, loud unexplained noises, or a sensed presence. "There have always been such fields," Persinger notes, but "they are not continuous. Instead they occur as brief transients that require patience and precision to record." Parapsychologists might say that this is because the presence, or ghost, comes and goes. When Persinger and Hill replicated the patterns in the lab, "most people report a sensed presence." The subjects were told that they were participating in a 'relaxation' study, so anticipation was minimal. But after putting on the helmet they said things like, "I see shadows along my left side . . . there is someone touching my left side . . . I see a visual . . . it's an apparition." One woman who felt a presence said, "I started to cry when I felt it slowly fade away." The experience was scarier when the right hemisphere was "preferentially stimulated," Persinger discovered. They'd report "a dark, ominous force looming right above" and "odd smells, terror and other classic haunt patterns," like flickering images of skeletons. Apparitions that appeared to someone's left side were usually reported as frightening, but apparitions on the right were often attributed to a dead relative, an angel, or Christ.

Persinger investigated one house where a young couple had reported an "apparition moving through their bed" and also the sound of breathing. What he found there were repeated transient magnetic field patterns that were "similar to those that evoke the sense presence in our experimental studies." Persinger also investigated a young girl who was seeing apparitions of a baby over her left shoulder. "Direct measurement around her bed showed the occurrence of a pulsed magnetic field whose structure was similar to those we employ to evoke the sensed presence in the laboratory.

"Many of the patterns that are found in nature, including 'haunted areas,'" Persinger points out, "are also generated by the human brain during altered states or during conditions that are similar to complex partial seizures without convulsions. This might help explain why some people are more sensitive than others in haunted areas. Individuals with temporal lobe sensitivity, that can be measured by psychometric indicators, respond very powerfully to these weak fields," Persinger discovered. He also found that people who have sustained mild brain injury and describe themselves as "not the same people as they were before the injury" had more paranormal experiences, like feeling the presence of a sentient being. Mediums may either have temporal lobe sensitivity or an injury to the brain. Peter Hurkos always said his abilities came after a fall off a ladder.

When asked if there was any possibility that the EMF fluctuations in the field might represent an intelligent presence, Persinger answered, "Stan Koren and I wrote a chapter for Houran and Lange (*Hauntings and Poltergeists*, 2001) where we discuss the possibility that configured magnetic flux lines within a small space (such as the luminosities seen in haunt and in ghost light areas) might have the energy density, intraspatial complexity (e.g., similar to trillions of synapses in the human brain) and timing to allow 'intelligence' to emerge, at least transiently. Of course this intelligence could be suspended, just like ours when we enter deep sleep at night, only to return when the functional reconfiguration occurs again. This is a hypothesis well worth pursuing." In an earlier paper he wrote, "there is some evidence that some paranormal experiences may be transformations of information not normally accessible." The answer appears to be "maybe." A fleeting apparition or Raymond Bayless's Spook Light might simply be an intermittent signal, or information, available only to those with the proper tuning or filtering mechanism.

In the end, it's frustrating how uncommunicative apparitions appear to be. In general they never seem to convey substantive information, like what it's like to be dead. Hans Holzer says it's because only the most damaged people stay behind. Auerbach says it's because they don't know. They are dead people who are still here. One apparition told Auerbach that she was afraid of going to hell, so when she died she concentrated on being back home, and all of a sudden she was. She never saw a light to go toward. One minute she was in the hospital dying and the next minute she was back home. Dead, but still there. Therefore she doesn't have a lot to say about the other side because she hasn't really been there yet. But she can and does talk about what it's like to no longer have a physical body. She described herself as "a ball of energy," without form, and she was seen when she projected her appearance into the minds of others.

It sounds much like what the lab said was the key to survival and life after death all those decades ago: telepathy.

ELEVEN

Just before Rhine retired from Duke in 1965, a seventy-five-year-old inventor named Glenn W. Watson started writing Clement Stone, one of Rhine's financial contributors. He was looking for financing for Telepathy-Type, a typewriter that would type out messages received telepathically. Watson, the inventor of the radio typewriter, had come up with the idea thirty years before. AT&T had worked for years on a similar project, Watson claimed, but had gotten nowhere. Stone asked Rhine what he thought, and Rhine said the idea had no merit. Engineers young and old had attempted approaching the problem this way, he said. But what they all failed to understand was "that the limiting factor is in the individual, in the deep recesses of human personality, and not in the gadgetry of transmission." Watson's plan was "based on a complete misconception of what telepathy is and how it works."

Point taken about the limiting factor. But no one really knew what telepathy was or how it worked, including Rhine.

All he did was establish that it was there, and to this day, many would argue that he hadn't accomplished even that. This would be made miserably clear in 1960, at the end of the third and final panel discussion at a Dartmouth College conference about consciousness, when Dr. Warren Weaver rose and asked to be recognized by the chair.

"I had rather hoped that this would be introduced by one of my youngers and betters," Weaver began, "but since no one else has, I'm going to mention what is obviously a controversial topic . . . I am in fact referring to that embarrassing, partially disreputable but nevertheless challenging body of phenomena known as extrasensory perception."

Then he said something that went too far. "I would like to mention the fact that I find this whole field intellectually a very painful one. And I find it painful essentially for the following reasons: I cannot reject the evidence and I cannot accept the conclusions." When Dr. Weaver suggested that this would be an excellent topic the next time Dartmouth arranged a convocation, the chair of the session rather brutally cut him off and would not let him continue. A few people came up to Weaver afterward to say how glad they were that Weaver had raised the issue, including the president of the college, John Sloan Dickey, but the majority of the crowd sided with the chairman, and later, at another conference in Minneapolis a colleague of Weaver's rushed up to him to tell him that their colleagues had expressed the fear that "Weaver has gone off his rocker."

Weaver was a former president of the AAAS, a man who over his career directed the allocation of hundreds of millions of dollars in grants in science and medicine, and who co-authored with Claude Shannon *The Mathematical Theory of Communication*, a milestone in communication theory. His ideas about machine translation would influence some of the most important computer pioneers of his time. The Computer Science Department of New York University is housed within Warren Weaver Hall. But when he got up to merely suggest that there was something worth investigating in parapsychology, for the people in that audience it was the equivalent of Bill Gates getting up in front of a computer association and saying, "Our tests seem to indicate computers run on fairy dust." Even someone of Weaver's stature couldn't change the fact that all those years ago when Rhine faced down the critics at the APA conference in Columbus, it was not the victory he thought it was.

A few years after Weaver confronted the audience at Dartmouth, the British researcher Mark Hansel published his theory that Hubert Pearce had cheated. According to Hansel, Hubert had left the library and stood on a chair in a room across the hall from Gaither. Hansel then proposed that Hubert was able to look through the transom window, across the hall, through a second transom window and down to the cards that Gaither had turned over. The rooms Hansel saw had been altered since the time of the experiments, but blueprints of the original layout showed that cheating in this way would have been physically impossible; there was no line of sight. Apparently Hansel did not accept the diagram. Plus he had a backup theory. Maybe Pearce crawled through the ceiling and peeked down. Without Pratt hearing him. Or anyone else seeing him. Over thirty times. Through a trapdoor that didn't exist.

The fact that more than three decades later Pearce's results were still being picked apart was enormously discouraging.

A lot of their old friends were gone by now. One of their most faithful contributors, Charles Ozanne, had started showing signs of dementia and was found wandering lost and alone in a cemetery one night. He died two years later, in 1961. Upton Sinclair's wife, Mary Craig, died a few weeks after him on April 26. It was her experiments that made McDougall charge Rhine with the goal of proving telepathy, the first step toward proving life after death. But when Mary Craig finally died, she left this world absolutely terrified and unsure about the afterlife and without any definitive answers on the matter from Rhine. "Her fears dominate her whole being," Upton wrote his son.

Surprisingly, while Mary Craig's death left Upton with a "sense of desolation beyond my power to describe," he didn't deal with his desperate loneliness by reaching back to the now deceased Mary Craig. There are no records of any séances. Arthur Ford, the medium both Mary Craig and Upton had so much faith in, and who in a few years would help Bishop Pike communicate with his dead son, was never consulted. Six months after Mary Craig died Upton Sinclair remarried.

Duke University had initially planned to establish what they called the Center for the Study of the Nature of Man after Rhine retired. It wasn't going to be just a parapsychological institute, but a collaborative effort involving parapsychology, psychology, physics, electronic engineering, certain medical branches, statistics, philosophy, literature, history, anthropology, and religion. But Rhine was concerned that parapsychology was going to be subsumed by Duke's center, and Duke University officials were having second thoughts about the whole thing after having confidential conversations with Karl Zener and roughly a half dozen other senior professors at Duke. Duke eventually dropped their plans for a center, and Rhine set up the Foundation for Research on the Nature of Man (FRNM, pronounced FURnum) independently of Duke on August 1, 1962.

There was one problem. It had always been understood that Gaither Pratt would take over when Rhine retired, and now, not only would there be no lab to take over, there would no longer be a place for Gaither at Duke. Rhine planned to take Gaither with him to FRNM, of course, but while Rhine would retire from Duke with all the benefits and privileges of a full professor, Gaither would not. Over the years, Gaither had been offered positions at Duke and other places that would have given him the same benefits, but to take them meant giving up parapsychology and that was unthinkable. Everyone at the lab imagined that parapsychology would cause a scientific revolution like the discoveries of Copernicus, Newton, or Einstein, and unfortunately there was no financial security in that. In 1963 there was a messy breakup between Rhine and Gaither, and the following year Pratt was working at the University of Virginia Medical School, where Ian Stevenson, a professor and the chairman of the Department of Psychiatry, was conducting parapsychological research into reincarnation and other areas more directly related to life after death.

Two months after the Parapsychology Laboratory of Duke University closed its doors, the Broadway play *On a Clear Day You Can See Forever* opened at the Mark Hellinger Theatre in New York City. The play is about a woman with psychic abilities who learns through hypnosis that she has been reincarnated. Harvard graduate Alan Jay Lerner, who wrote the libretto and the lyrics, had been studying ESP for years. "The weight of evidence is that we all have a vast latent extrasensory perception," he told a *New York Times* reporter. His lyrics reflected a growing trend in the country and in parapsychology toward the individual and self-actualization. The songs had lines like "So much more than we ever knew, So much more were we born to do," and "the glow of your being, Outshines every star."

When Rhine and his staff made speeches the year before about their plans for the future, the theme was the same. There was no talk of love, death, loss, and survival. The emphasis had moved toward understanding man and what he could do now and not hereafter. What powers are within us? How can we be more than what we are? "There has never been a greater need to know what man has within his nature for use in self-understanding," Rhine said. Parapsychology had evolved—it was not about reconnecting with what was dead and gone, but with the "hidden part of man's nature," the "unknown side of human personality represented by psi." Finishing his speech with a glorious call to arms, Rhine said that science would "be enabled, after all its sweeping conquests of the outer worlds of space and time, of computers and molecules, of galaxies and microparticles, to catch up at last with man himself." The fact that science had not accepted their results up until this point was not mentioned.

Initially, the changeover from Duke to FRNM had been in name only. They were still working out of the West Duke Building, so all they did was replace the stationery and the door sign. Later in December they moved across the street and into the home of a former dean where Rhine could still see the West Duke Building from his office window. There had been some talk about leaving Durham altogether, but after a flurry of editorials and meetings with university officials and the governor, who brought up Roosevelt's response to Einstein's proposal for the Manhattan Project, they decided to stay where they were.

The year 1965 got off to a good start. Rhine joined a lecture bureau and went on an eleven-city tour that was reminiscent of his first eleven-city tour in 1936, when he traveled across America to introduce everyone to ESP. Only this time he was paid up to seven hundred fifty dollars per engagement and everyone knew what ESP was. He and Gaither had patched things up and were once again talking like people who had known and respected each other their whole lives. One of the first things Gaither did when he got to the University of Virginia was resume work with Hubert Pearce. "We gave up too early and too easily in the Duke Laboratory on the possibility of selecting really good subjects for our research," Gaither wrote his old friend. "I am eager to attempt to get back to the point at which we got off trail." Gaither was now fifty-five years old and Hubert was sixty, but Hubert was ready to give it another go and they would try repeatedly. Gaither even tested one of Hubert's sons, but he didn't score above chance. Every time Rhine or Gaither asked, Hubert made himself available for tests. But he never produced astounding results again. In eight more years he'd be dead.

Rhine's focus now was on the young. In the fall of 1965 they had a part-time relationship with more Duke graduate students than they had had in some time. Unburdened with the usual baggage, the new students were parapsychology's clean slate, their hope for the future. The subject was as alluring as ever to them, and Rhine still knew how to beguile the young. "He could create an atmosphere just by being part of it," one said. This was the sixties, and the newly rebellious young people of America were even more attracted by the outsider status and self-empowering promise of parapsychology. After spending time with Rhine, "the idea we came away with," the wife of a staff member remembers, "was that parapsychology was a dangerous, romantic adventure." That a man who was now in his seventies could accomplish this with a group in their twenties and thirties said something about the power of his charisma.

But Rhine was still unwilling to deviate from the parapsychological path they had taken more than thirty years before. When one of his younger workers wanted to try something new the answer was invariably no. After thirty years of trying they still didn't have a working theory of ESP, but the idea of throwing in the towel now and taking a completely different course was unthinkable. It would be like saying all those years were a waste. Besides, Rhine believed they were almost there so they weren't going to depart from their course now. Not when they were *this* close.

The sixties ultimately worked against Rhine. In addition to everything else, these were antiauthoritarian times, and by 1967 Rhine lost the group of young people he had so carefully selected. A lifetime of professional disdain had tweaked his already determined (stubborn) nature into a full-blown bullheadedness that bordered on paranoia. When Rhine fired one of their most promising members, the rest of the group immediately resigned, leaving Rhine and FRNM with only Louie, his daughter Sally, and one other researcher, who left two years later.

But just as the sixties were about to end they had an important and long hoped for triumph. It was 1969 and the Parapsychological Association was making its fourth attempt to become an affiliate member of the prestigious AAAS. On December 30, 1969, in Boston, the AAAS Council prepared to vote. The American Orthopsychiatric Association was considered first. It was almost perfunctory. Within seconds they were voted in. "Now the Parapsychological Association," the chairman called out. "Do I hear a motion?" Ten seconds of silence. "Yes," one lone voice called out. This was followed by another excruciating wait before the motion was seconded. But there was no pause before people started raising their hands to object. "The so-called phenomena of parapsychology do not exist and it is impossible to do scientific work in this area," one scientist called out. "We are not familiar with what para-psychology is and so we are not qualified to make a vote," said another.

They'd been voted down again and again, and it looked like they were going to be voted down this time too. Then famed anthropologist Dr. Margaret Mead stood up. "For the past 10 years, we have been arguing about what constitutes science and scientific method and what societies use it," she began. Blinds, double-blinds, statistics—the parapsychologists use them all, she pointed out. "The whole history of scientific advance is full of scientists investigating phenomena that the establishment did not believe were there," she said. "I submit that we vote in favor of this association's work." It was time, the chairman declared. "Please raise your hands those Council members in favor of the motion." Everyone looked around the room, ready to start counting. There was no need. Most of the people in the room had their hands in the air. "Those against?" A decent number put up their hands, but they were still obviously a minority. The motion carried. The Parapsychological Association was accepted. Douglas Dean, who had put together the application and all the packages of supporting materials and had run around making sure everything was in order, couldn't stop himself from crying. They had finally arrived.

But old friends continued to disappear. While everyone had been preparing for the AAAS vote, Eileen Garrett was in so much pain from osteoarthritis she had to be driven the few short blocks from her apartment to the Parapsychology Foundation office. Once there her daughter would inject her with enough painkillers to get her through the day. Allan Angoff, who had been with the foundation for years wrote, "she sometimes wept, sobbed for her disability and for what she knew was the inexorable ending of the excitement she so enjoyed." She somehow still made it to Nice in June 1970 to participate in a Parapsychology Foundation conference, but she would never leave that city. She died there on September 15, at age seventy-seven, from a stroke, and without ever learning for certain what the source of her unusual abilities was.

Rhine was now seventy-four years old and needed someone to replace him at FRNM, but there was a problem with every candidate, including his own daughter Sally. She tried running the center her way, but Rhine had been the boss for too long. Helmut Schmidt, a physicist at Boeing Aircraft, who had developed the first experiments to investigate psychokinesis using random number generators, was given a job and considered next. That didn't work out either, and Schmidt finds the memories so hurtful he refuses to talk about it. Rhine's next choice resigned a year later following an allegation that he falsified records in 1974.

A couple of years later a new group called the Committee for the Scientific Investigation of Claims of the Paranormal (CSICOP) had formed to take away what little ground Rhine and his colleagues had gained. Not surprisingly, B. F. Skinner was associated with CSICOP. Dr. Marcello Truzzi, who had helped start CSICOP's crusade, split with them a couple of years later. "I don't doubt that 99 percent of occultism is empirically false," he explained, "but the approach to it has to be based on an examination of the evidence by people qualified to do that, not an outright condemnation." The work of Rhine and others was healthy, he said. They were investigating "matters that present legitimate puzzle areas for science." Skepticism "refers to doubt rather than denial," Truzzi wrote, and "critics who take the negative rather than an agnostic position but still call themselves 'skeptics' are actually pseudo-skeptics. . . the true skeptic takes an agnostic position, one that says the claim is not proved rather than disproved."

When Rhine appointed the philosopher and psychologist K. Ramakrishna Rao as director in 1977 he decided their next best course of action was to return to Duke. He first broached the subject with Duke President Terry Sanford, gently referring to it as a possible affiliation. The following year rumors circulated that Princeton University was setting up some sort of parapsychology research center. Rhine wrote Sanford again. "I am not sure it was either good or necessary to move it [the lab] off the campus with my retirement," he confided. In any case, "you were not President then," and things have changed. But nothing happened. The next year, 1979, with financial support from James McDonnell of McDonnell Douglas, the Princeton Engineering Anomalies Research (PEAR) program was established by Robert G. Jahn, then dean of the School of Engineering and Applied Science. A bequest in Chester Carlson's will had already enabled the University of Virginia to establish the Division of Parapsychology. The man who brought parapsychology to academia was suddenly, at the end of his life, on the outside looking in.

Rhine submitted a proposal to the new provost the following March. The provost said he would take it up with the chancellor, but Rhine's health was rapidly failing.

While Rhine was fighting for his life's work and his life, the physicist John Archibald Wheeler, who had worked with Einstein, got up and addressed the AAAS in Houston at a panel called "The Role of Consciousness in the Physical World," organized by Robert Jahn. Wheeler was dismayed at having to share the podium with parapsychologists and he wanted to say something about it. Like Rhine when he addressed the APA conference in 1938, Wheeler was clearly nervous. Before he began he stiffened, his face turned red, and his speech quickened and went up in register. "There is nothing that one can't research the hell out of," he said. "Research guided by bad judgment is a black hole for good money." Wheeler was, in fact, the scientist responsible for coining the term "black hole." "Where there is meat there are flies," he heartlessly continued. "Now is the time for everyone who believes in the rule of reason to speak up against pathological science and its purveyors." He wanted the Parapsychological Association thrown out of the AAAS. When the audience asked him to be more specific about his criticisms of parapsychology he accused Rhine of falsifying an experiment fifty years before and said he had a witness. The witness later wrote and completely rejected the charges, and Wheeler's retraction, described as "grudging and meager," was published in the AAAS magazine *Science* the following summer. Once again, if Rhine thought their acceptance into the AAAS indicated that they were finally getting somewhere he was as wrong as he had been after the APA conference in Columbus all those years ago.

By now Rhine's eyesight and hearing were almost gone, along with his sense of taste and smell, but he remained upbeat. On August 21, 1979, Rhine wrote Gaither, "life is still great, even with all these problems and puzzles." Gaither was now retired. One of their earliest critics had once said, "the energies of young men and women in their most vital years of professional training are being diverted into a side-issue." But Gaither had no regrets. He died on November 3, 1979, in Charlottesville. He was sixty-nine.

Rhine wrote Sanford the next month to see if he would accept a position on FRNM's board of directors. He no doubt saw this as a step toward the affiliation he sought. He waited for Sanford's response. By Christmas Rhine was blind and deaf. Two months later, on February 20, 1980, at the age of eighty-four, J. B. Rhine was dead. The ESP breakthrough that he surely hoped would occur during his lifetime never arrived. Neither did the return to Duke. The University of North Carolina at Chapel Hill was approached after Rhine's death, but they declined as well.

As a young woman just starting out in life and marriage, Louie had imagined her husband's death and had written, "I hope I won't have to stay long behind alone." She lived on for what may have been for her four long years. In 1983, right after completing a highly personal book called *Something Hidden*, Louisa Rhine died. She was ninety-one.

In 1996 Duke University's campus newspaper *The Chronicle* printed a letter once again challenging the work of J. B. Rhine and the Parapsychology Laboratory.

A number of people wrote in to defend him, including Gaither Pratt's son Vernon, who said, "Much is not known about the origin and causes of parapsychology, but many occurrences not explainable by known causes have been witnessed responsibly, proven beyond probability and properly recorded. They can be ignored, but they cannot be wished away."

At the very beginning of their careers the Rhines wrote in their report about the medium Mina Crandon, "If we can never know to a relative certainty that there was no trickery possible, no inconsistencies present, and no normal action occurring, we can never have a science and never really know anything about psychic phenomena." Did they ever look back at these words at the end of their lives and feel pity for the long-dead medium?

Because many would sum up their work almost the same. "Scientists like Dr. J. B. Rhine of Duke University have not labored all these years in parapsychology to prove merely that the human mind can correctly guess more than one out of five test cards," Morey Bernstein once said. But even though some predicted Rhine's work would change everything, his work has since been largely ignored. While Gaither Pratt and others prematurely announced that the battle was won, that "parapsychology is here to stay. The time of testing by ordeal in the fire of scientific scorn has passed," the truth is that as of now, it has been lost.

Three quarters of a century ago the scientists at Duke University repeatedly demonstrated the existence of a group of anomalies that they chose to call extrasensory perception. Since then, the scientific community has failed to either convincingly disprove them or come up with a plausible alternative explanation. Unfortunately, the field of parapsychology itself has failed to shed any definitive light on them either.

Stalemate.

Explanations may one day be found that will either make it all go away or increase our understanding, but for now, the results sit in the Special Collections Library of Duke University, ignored for decades, and the researchers who devoted their careers to find one piece of incontrovertible evidence out of centuries of hope, fear, superstition, and supposition are called liars, dupes, and incompetents when they are remembered at all.

Very few people can know if what they do in the short time they're given will make a significant contribution to humanity, to history, or to anything at all. What looks significant today can disappear in a moment, supplanted by an even better idea; what is ignored now may turn out to be the key to a revolution. There are many who believe the Rhines will one day be hailed as historic figures, but it looks like that won't be sorted out until anyone reading this is gone.

In 1930 Einstein said, "The most beautiful thing we can experience is the mysterious. It is the source of all true art and science." A half century later Betty McMahan wrote that the phenomena they studied were "tantalizing sparks thrown out by the universe—hints of how it is put together. Hints that should be followed up in our intense desire to clarify the magnificent mysteries of existence . . . every piece of the puzzle is important, and eventually will join with other pieces, provided by other disciplines, to reach the final solution." Progress could be made as long as we used all the science at our disposal, she went on, and followed every serious field of inquiry, whether it was through the studies of quantum physics or parapsychology. Recent overlap between information theory, quantum mechanics, theories about consciousness, and parapsychology have produced still more tantalizing sparks. In 2001 Nobel Laureate Brian Josephson wrote, "Quantum theory is now being fruitfully combined with theories of information and computation. These developments may lead to an explanation of processes still not understood within conventional science such as telepathy." Others have introduced theories about how consciousness and nonlocality may operate within the brain to explain telepathy. But we are not there yet. The work done at Duke could some day change everything. Or it could be added to the list of scientific ideas that seemed like a good idea at the time.

Dorothy Pope said that if you were going into parapsychology, "the only safe thing to do is put your main reliance on your devotion to the field, the sense of satisfaction that you get out of being a pioneer." In spite of everything, it had its rewards. "Every day was a great day. We were adventuring, experiencing, and learning. We were serious workers, but there was that about our work that really made it like exciting play."

William James said, "In psychology, physiology and medicine, whenever a debate between the mystics and the scientists has been once and for all decided, it is the mystics who have usually proved to be right about the facts, while the scientists had the better of it in respect to the theories." The survival debate has not been decided. The story of the Parapsychology Laboratory begins and ends in stalemate. Their experiments confirmed telepathy and were never generally accepted. They looked for evidence of life after death, but the evidence was inconclusive. Everyone we love dies and disappears, and whether or not something more substantial than a memory survives of all that love is, scientifically, an unanswered question. In the end the men and women of the Parapsychology Laboratory were left with what Gaither described as "an uneasy truce between the dread of the hereafter and the solace of faith," which could, more than likely, be a truce we'll have to hold on to forever.

EPILOGUE

It was the parapsychology critics themselves who finally convinced me that the lab's work was sound. I followed decades' worth of debates and found that after hearing the lab's answer to their objections, critics either withdrew their complaints, graciously conceding that their objection had been without merit or had since been adequately addressed. Or they kept silent, with the ball in their court, neither conceding nor continuing the debate, not wanting to credit Rhine with having satisfactorily answered their objections.

There were two notable exceptions: Mark Hansel and William Feller. Hansel kept coming up with more arguments, some so obviously biased and far-fetched (the trapdoor theory) it's hard to understand how they ever got published in the first place.

The more problematic exception was William Feller. Feller was a Princeton mathematician who specialized in probability theory and had enormous stature in the field of statistics. Unlike Hansel, he cannot be easily dismissed. Among the problems he had with the lab's work was their methods of shuffling, and the belief that they practiced optional stopping (ending the experiment while they were still getting good results). The lab responded to each of his points, but Feller never conceded and never accepted their results.

Persi Diaconis, a former magician and now a professor of statistics at Stanford University (and a former student and great admirer of the mathematician), looked at Feller's criticisms and found that several were wrong. But after talking to students and colleagues of Feller's, he discovered his "mistakes were widely known," and concluded that Feller "seemed to have decided the opposition was wrong and that was that." It was like Einstein's remark to his secretary, "Even if I saw a ghost I wouldn't believe it."

Rhine's colleague Don Adams, who was one of many who called for Rhine to be restrained (or worse), later expressed remorse for his actions in an essay called "The Natural History of a Prejudice." Adams admitted that even though Rhine's statistics "seemed impeccable and his gradually more rigorous conditions adequate," he nonetheless longed for Rhine to fail. "I wanted not the truth, but to prove his positive conclusions wrong."

I came across this bias again and again. What do these scientists have to lose if Rhine were to be proven correct? The very scientific method that they live by requires that they abandon tenets, however well fortified, when better data are presented. You'd think that scientists—who throughout time have introduced sometimes unimaginable discoveries and endured unreasonable and occasionally lethal resistance—would be more amenable. But physicists initially argued against evolution because it didn't fit their view of the physical world at the time. In order to incorporate Darwin's discovery cherished views had to be sacrificed. The kind of damage a new discovery inflicts is not merely intellectual, but emotional. When looking back at his behavior toward Rhine, Don Adams wrote, "Have you ever had the experience of seeing a belief, that you have considered fundamental to everything you value, gradually but inexorably undermined? . . . I have never had much sympathy for the embattled Fundamentalists, but since facing a situation comparable in many ways to theirs and finding that I behaved just as badly, their conduct no longer seems so strange."

But science is a work in progress, and the guarantee of still more fantastic discoveries and subsequent damage is a given. Knowledge will advance. The lab's data weren't quite enough to overhaul existing hypotheses, but for those willing to study them dispassionately, these yet-to-be-accepted effects, or anomalies, which Rhine called telepathy and psychokinesis, might one day provide a valuable scientific clue to the nature of the universe. "Is it possible," Stanford University physicist Andrei Linde asks, "that consciousness, like space-time, has its own intrinsic degrees of freedom, and that neglecting these will lead to a description of the universe that is fundamentally incomplete?" Are the properties of consciousness an important piece of the cosmological puzzle?

The work of the Duke Parapsychology Lab will probably be explained some day, one way or another. Critics are sure that time will show that it was all the result of poor technique or some yet to be discovered flaw in the math, and maybe it will. But we don't know that. For now that is only a belief—faith, not science. Until that time, both sides are holding fast. For the public, Rhine's work holds out the possibility that ghosts are real and love lasts forever, and they're not going to let go of that glimmer of hope any sooner than his critics are going to concede that Rhine just might have been onto something.

In the end, parapsychology's critics couldn't come up with convincing evidence to reject the lab's work and neither could I. But I did take away two ideas: There might be an unidentified source of information out there, along with unknown methods of transmission and processing. Also, we have a lot more to learn about consciousness.

■

The Parapsychology Laboratory, which became the Foundation for Research on the Nature of Man, is now called The Rhine Center: An Institute for Consciousness Research and Education. J.B. and Louie's daughter Sally Rhine Feather is the executive director. The Parapsychological Association, which Rhine conceived and help found, is still an affiliate member of the AAAS.

The Parapsychology Foundation also continues today. Eileen Garrett's daughter, Eileen Coly, is the president, and her granddaughter, Lisette Coly, is the executive director and editor of their publication, the *International Journal of Parapsychology*.

Bill Roll continues to direct the Psychical Research Foundation, which is now located in Carrollton, Georgia.

The University of Virginia Medical School's Division of Parapsychology, where Gaither Pratt went to work for Ian Stevenson, is now the Division of Perceptual Studies and is part of the University of Virginia's Department of Psychiatric Medicine. Ian Stevenson died in 2007, and Dr. Bruce Greyson has been the director since 2002. According to Jim Tucker, who specializes in reincarnation cases at the division, they now have twenty-five hundred cases of reincarnation.

The Princeton Engineering Anomalies Research (PEAR) program closed in 2007.

SOURCES

A good deal of this book came from letters and materials from the Parapsychology Laboratory Records, 1893–1984, Rare Book, Manuscript, and Special Collections Library, Duke University, Durham, North Carolina. Ultimately thousands of articles from that collection inform this book; I cite only materials I quote from directly.

EPIGRAPH

Aniela Jaffe, "The Psychic World of C. G. Jung," *Tomorrow*, Spring 1961.

CHAPTER ONE

Interviews with or information supplied by Carlos S. Alvarado, Elizabeth McMahan, James G. Matlock, Robert Rhine, Sally Rhine Feather, Anna Thurlow, Nancy L. Zingrone.

Clement Cheroux, Andreas Fischer, Pierre Apraxine, Denis Canguilhem, Sophie Schmit, *The Perfect Medium: Photography and the Occult*, Yale University Press, 2004.

Eileen Coly, *Eileen J. Garrett: Adventures in the Supernormal*, Helix Press, 2002.

Robert F. Durden, *The Launching of Duke University 1924–1949*, Duke University Press, 1993.

Seymour H. Mauskopf and Michael R. McVaugh, *The Elusive Science: Origins of Experimental Psychical Research*, Johns Hopkins University Press, 1980.

Louisa E. Rhine, *Something Hidden*, McFarland & Company, 1983.

Thomas R. Tietze, *Margery*, Harper and Row, 1973.

"The Margery Mediumship," *Boston Evening Transcript*, February 18, 1925.

"Margery Genuine, Says Conan Doyle; He Scores Houdini," *Boston Herald*, January 26, 1926.

"Ectoplasm Prints Called Lung Tissue," *New York Times*, February 28, 1926.

J. B. Rhine, Ph.D., and Louisa E. Rhine, Ph.D., "One Evening's Observation on the Margery Mediumship," *Journal of Abnormal and Social Psychology*, vol. 21, no. 4, January–March 1927.

"Declared Tricksters by Former Partisans," *Boston American*, February 7, 1927.

"Doyle Declares Margery Genuine," *New York Times*, March 4, 1927.

John F. Thomas, Ph.D., "Beyond Normal Cognition: Being an Evaluation and Methodological Study of the Mental Content of Certain Trance Phenomena," Ph.D. dissertation, Duke University, March 1937. Published by Boston Society of Psychical Research.

Edmond P. Gibson, "The Ethel Thomas Case," *Tomorrow*, vol. 2, no. 4, Summer 1954.

Louisa E. Rhine, "J. B. Rhine: Man and Scientist," from *J. B. Rhine: On the Frontiers of Science*, ed. K. Ramakrishna Rao, McFarland & Company, 1982.

James G. Matlock, "A Cat's Paw: Margery and the Rhines, 1926," *Journal of Parapsychology*, vol. 51, September 1987.

Taped interview of J. B. and Louisa Rhine, by Seymour H. Mauskopf and Michael R. McVaugh, 1974, Seymour H. Mauskopf Papers, 1972–1982, Duke University Special Collections.

Letters and materials from the Parapsychology Laboratory Records, 1893–1984, Rare Book, Manuscript, and Special Collections Library, Duke University, Durham, North Carolina:

Grant Code to Walter Prince, 1926.

J. B. Rhine to John Thomas, August 15, 1927.

J. B. Rhine to Dr. Prince, August 15, 1927.

J. B. Rhine to John Thomas, November 10, 1927.

Nandor Fodor to J. B. Rhine, April 12, 1944.

CHAPTER TWO

Interviews with or information supplied by Eileen Coly, Lisette Coly, Elizabeth McMahan, Warren Pearce, Ellen Pratt, Robert Rhine, Sally Rhine Feather, Rhea A. White, Nancy L. Zingrone.

Denis Brian, *Einstein: A Life*, J. Wiley, 1996.

Eileen Coly, *Eileen J. Garrett: Adventures in the Supernormal*, Helix Press, 2002.

Seymour H. Mauskopf and Michael R. McVaugh, *The Elusive Science: Origins of Experimental Psychical Research*, Johns Hopkins University Press, 1980.

Elizabeth Lloyd Mayer, Ph.D., *Extraordinary Knowing: Science, Skepticism, and the Inexplicable Powers of the Human Mind*, Bantam Books, 2007.

J. G. Pratt, J. B. Rhine, Burke M. Smith, Charles E. Stuart, Joseph A. Greenwood, *Extra-Sensory Perception After Sixty Years*, Henry Holt and Company, 1940.

J. B. Rhine, *Extra-Sensory Perception*, Faber and Faber, 1935.

Louisa E. Rhine, *Something Hidden*, McFarland & Company, 1983.

Upton Sinclair, *Mental Radio*, Hampton Roads Publishing, 2001 (originally copyrighted 1930 by Upton Sinclair).

"Rhine's 289 Wins President's Match," *New York Times*, August 22, 1919.

Austin C. Lescarboura, "Edison's Views on Life and Death as Reported," *Scientific American*, October 30, 1920.

Dr. William McDougall's letter to the editor, *New York Times*, May 27, 1928.

"Prof. Einstein Begins His Work at Mt. Wilson; Hoping to Solve Problems Touching Relativity," *New York Times*, January 3, 1931.

"Einstein Drops Idea of Closed Universe," *New York Times*, February 4, 1931.

J. B. Rhine, "Telepathy and Clairvoyance in a Trance Medium," *Scientific American*, July 1935.

Francis Sill Wickware, "Dr. Rhine & E. S. P.," *Life*, April 15, 1940.

Margaret Pegram Reeves and J. B. Rhine, "Exceptional Scores in ESP Tests and the Conditions I. The Case of Lillian," *Journal of Parapsychology*, vol. 6, no. 3, September 1942.

Edmond P. Gibson, "The Ethel Thomas Case," *Tomorrow*, vol. 2, no. 4, Summer 1954.

J. B. Rhine and J. G. Pratt, "A Review of the Pearce-Pratt Distance Series of ESP Tests," *Journal of Parapsychology*, vol. 18 , no. 3, September 1954.

Aniela Jaffe, "The Psychic World of C. G. Jung," *Tomorrow*, Spring 1961.

Albert Einstein to Jan Ehrenwald, May 13, 1946, *The Zetetic*, vol. 2, no. 1, Fall/Winter 1977.

Albert Einstein to Jan Ehrenwald, July 8, 1946, *The Zetetic*, vol. 2, no. 2, Spring/Summer 1978.

Kent Marts, "ESP Preacher Dared Not Talk About a Feeling," Benton County *Democrat*, September 14, 1986.

Rhea A. White, "The Narrative Is the Thing: The Story of 'Necessary Spirit' and Psi," *Journal of the American Society for Psychical Research*, vol. 96, July/October 2002.

George Pendle, "Einstein's Close Encounter," (London) *Guardian*, July 14, 2005.

From the Sinclair Mss., Manuscripts Department, The Lilly Library, Indiana University Bloomington:

William McDougall to Upton Sinclair, June 26, 1929.

Upton Sinclair to William McDougall, July 19, 1929.

William McDougall to Upton Sinclair, September 27, 1930.

Upton Sinclair to William McDougall, January 26, 1931.

Upton Sinclair to William McDougall, March 9, 1931.

Letters and materials from the Parapsychology Laboratory Records, 1893–1984, Rare Book, Manuscript, and Special Collections Library, Duke University, Durham, North Carolina:

Transcript of sitting with Mrs. Garrett, April 14, 1934, Gaither Pratt and Miss Crandall.

Transcript of sitting with Mrs. Garrett, April 16, 1934, Dr. Adams and Miss Crandall.

Transcript of sitting with Mrs. Garrett, April 16, 1934, Dr. Thomas, Miss Crandall, and Hubert Pearce.

Transcript of sitting with Mrs. Garrett, April 17, 1934, Charles Stuart and Miss Crandall.

Transcript of sitting with Mrs. Garrett, April 17, 1934, Hubert Pearce and Miss Crandall.

Transcript of sitting with Mrs. Garrett, April 18, 1934, Mrs. Few and Miss Crandall.

Drs. Lundholm, Zener, Adams to Professor McDougall, April 9, 1934.

Hubert Pearce to J. B. Rhine, November 7, 1934.

Carl Jung to J. B. Rhine, November 27, 1934.

J. B. Rhine to Carl Jung, December 20, 1934.

J. B. Rhine to Francis Bolton, January 25, 1935.

Hubert Pearce to J. B. Rhine, 1937.

J. B. Rhine to Upton Sinclair, July 21, 1953.

Eileen Garrett to Aldous Huxley, January 14, 1959, from the Parapsychology Foundation Library.

CHAPTER THREE

Interviews with or information supplied by James Carpenter, Alice Bell Kirby, Elizabeth McMahan, Gertrude Schmeidler, Michael Simonson (Assistant Archivist, Central Database of Shoah Victims' Names), Jessica Utts.

Chris Carter, *Parapsychology and the Skeptics*, Sterling House Publishers, 2007.

H. M. Collins, *Changing Order: Replication and Induction in Scientific Practice*, Sage Publications, 1955.

Martin Gardner, *Fads and Fallacies in the Name of Science*, Dover Publications, 1957.

J. G. Pratt, J. B. Rhine, Burke M. Smith, Charles E. Stuart, Joseph A. Greenwood, *Extra-Sensory Perception After Sixty Years*, Henry Holt and Company, 1940.

Louisa E. Rhine, *Something Hidden*, McFarland & Company, 1983.

Marcello Truzzi, "Discussion on the Reception of Unconventional Scientific Claims," in *AAAS Selected Symposium 25, The Reception of Unconventional Science by the Scientific Community*, Westview Press, 1979.

Nancy L. Zingrone, "From Text to Self: The Interplay of Criticism and Response in the History of Parapsychology," Ph.D. dissertation, University of Edinburgh, 2006.

B. F. Skinner, "Is Sense Necessary?" *Saturday Review of Literature*, October 9, 1937.

Waldemar Kaempffert, "The Duke Experiments in Extra-Sensory Perception," *New York Times*, October 10, 1937.

Chester E. Kellog, "New Evidence (?) for 'Extra-Sensory Perception,'" *Scientific Monthly*, vol. 45, no. 4, October 1937.

"Rhine's Data on Clairvoyance Win Support of Statisticians," *New York Times*, January 30, 1938.

"The ESP Symposium at the APA," *Journal of Parapsychology*, vol. 2, no. 4, December 1938.

Edward V. Huntington, "Is It Chance or ESP?" *American Scholar*, vol. 7, 1938.

T. N. E. Greville, "The Application of Mathematics to ESP Problems (Including Selection of Data)," a paper delivered at the APA ESP symposium held in Columbus, Ohio, 1938.

"Girl's Psychic Powers Amaze Professors Come to See Her," *Baton Rouge State Times*, November 16, 1938.

Ted H. Maloy, "Child Mystic on 'Off Day' for Reporter," *Baton Rouge State Times*, November 17, 1938.

"Shy Spirits," *Newsweek*, November 28, 1938.

Persi Diaconis, "Statistical Problems in ESP Research," *Science*, New Series, vol. 201, no. 4351, July 14, 1978.

Seymour H. Mauskopf and Michael R. McVaugh, "The Controversy over Statistics in Parapsychology, 1934–1938," in *AAAS Selected Symposium 25, The Reception of Unconventional Science by the Scientific Community*, Westview Press, 1979.

Jessica Utts, "Replication and Meta-analysis in Parapsychology," *Statistical Science*, vol. 6., no. 4, 1991.

Stephen Schwartz, "The Blind Protocol and Its Place in Consciousness Research," *Explore*, vol. 1, no. 4, July 2005.

Sir Ronald Aylmer Fisher to J. B. Rhine, August 20, 1934, The University of Adelaide, Special Collections, The Collected Papers of R. A. Fisher.

Letters and materials from the Parapsychology Laboratory Records, 1893–1984, Rare Book, Manuscript, and Special Collections Library, Duke University, Durham, North Carolina:

Gardner Murphy to J. B. Rhine, December 20, 1936.

E. F. McDonald Jr. to J. B. Rhine, July 27, 1937.

Edward V. Huntington to J. B. Rhine, September 28, 1937.

Edward Rumley to J. B. Rhine, October 5, 1937.

B. F. Skinner to J. B. Rhine, November 15, 1937.

J. B. Rhine to B. F. Skinner, November 18, 1937.

B. F. Skinner to J. B. Rhine, November 23, 1937.

J. B. Rhine to B. F. Skinner, November 27, 1937.

J. B. Rhine to Charles Ozanne, January 10, 1938.

J. B. Rhine to Dr. Thorton C. Fry, January 11, 1938.

Mrs. Shelby White to J. B. Rhine, July 8, 1938.

Thomas N. E. Greville to J. B. Rhine, July 9, 1938.

J. B. Rhine to W. P. Few, August 29, 1938.

Lilli Guggenheim to J. B. Rhine, September 19, 1938.

J. B. Rhine to Upton Sinclair, October 13, 1938.

Betty Walker to J. B. Rhine, letter containing the report titled "The Case of Alice Bell Kirby," November 1938.

CHAPTER FOUR

Interviews with or information supplied by Elizabeth McMahan, Sally Rhine Feather, Rhea White.

Jurgen Keil, ed., *Gaither Pratt: A Life for Parapsychology*, McFarland & Company, 1987.

Elizabeth A. McMahan, *Heart and Nerve and Sinew*, self-published, 1990.

J. Gaither Pratt, *Parapsychology: An Insider's View of ESP*, Doubleday, 1964.

Louisa E. Rhine, *Something Hidden*, McFarland & Company, 1983.

Carl Sagan, *The Demon-Haunted World: Science as a Candle in the Dark*, Ballantine Books, 1997.

Francis Sill Wickware, "Dr. Rhine & E. S. P.," *Life*, April 15, 1940.

J. B. Rhine and Louisa E. Rhine, "The Psychokinetic Effect: I. The First Experiment," *Journal of Parapsychology*, no. 1, March 1943.

"The Ouija Comes Back," *New York Times*, September 10, 1944.

Parapsychology Bulletin, no. 3, August 1946.

Parapsychology Bulletin, no. 4, November 1946.

The Archive (a Duke publication), December 1946.

Edmond P. Gibson, "The Ethel Thomas Case," *Tomorrow*, vol. 2, no. 4, Summer 1954.

Letters and materials from the Parapsychology Laboratory Records, 1893–1984, Rare Book, Manuscript, and Special Collections Library, Duke University, Durham, North Carolina:

"A Bold Adventure," a proposal by J. B. Rhine (undated).

J. B. Rhine to Mrs. William Wood, July 13, 1940.

Gaither Pratt to Dr. Jerome S. Bruner, December 12, 1941.

J. B. Rhine to R. L. Flowers and Dean W. H. Wannamaker, January 12, 1942.

"If the case for PK is as good as I think it is," From J. B. Rhine letter dated August 18, 1942.

J. B. Rhine to The Hon. Francis P. Bolton, July 21, 1944.

Louisa Rhine to Ella [Weckesser], March 9, 1945.

Harold A. Scharper to J. B. Rhine, November 9, 1946.

J. B. Rhine to Harold A. Scharper, November 15, 1946.

Dr. Margaret Mead to J. B. Rhine, December 26, 1946.

Extract from minutes of meeting of the Board of Trustees of Duke University, held at Durham, North Carolina, on Wednesday, November 26, 1947, dated December 6, 1947.

CHAPTER FIVE

Interviews with or information supplied by Fr. James Flint, OSB (St. Procopius Abbey), Ralph Miner, Mark Opsasnick, Dr. Andrew B. Newberg, Robert Rhine, Shawn C. Wilson (Library Public Services Manager, The Kinsey Institute for Research in Sex, Gender, and Reproduction), Father Kenneth York (Diocese of Belleville, IL).

Denis Brian, *The Enchanted Voyager*, Prentice-Hall, 1982.

Naomi A. Hintze and J. Gaither Pratt, *The Psychic Realm: What Can You Believe?* Random House, 1975.

Henry Ansgard Kelly, Ph.D., *The Devil, Demonology and Witchcraft: The Development of Christian Beliefs in Evil Spirits*, Doubleday, 1968.

Kathleen R. Sands, *Demon Possession in Elizabethan England*, Praeger Publishers, 2004.

Troy Taylor, *The Devil Came to St. Louis: The True Story of the 1949 Exorcism*, Whitechapel Productions Press, 2006.

Alex Brett, Ph.D., and George M. Haslerud, Ph.D., "Bewitched Virginia Child Is Found to Be Normal," *Science News Letter*, January 14, 1939.

Gertrude Berger, "The Ouija Comes Back," *New York Times*, September 10, 1944.

J. B. Rhine, "The Relationship Between Psychology and Religion," *Carolina Quarterly*, June 1949.

Parapsychology Bulletin, no. 15, August 1949.

Western Folklore, "Folklore in the News," vol. 8, no. 4, October 1949.

K. E. Bates and Marietta Newton, "An Experimental Study of ESP Capacity in Mental Patients," *Journal of Parapsychology*, no. 3, December 1951.

Edward B. Fiske, "'Exorcist' Adds Problems for Catholic Clergymen," *New York Times*, January 28, 1974.

Robert D. McFadden, "Infant Boy Dies from Burns Laid to Exorcism Rite," *New York Times*, January 7, 1980.

Chuck Connors, "Personalities," *Washington Post*, May 11, 1988.

Mark Opsasnick, "The Haunted Boy of Cottage City: The Cold Hard Facts Behind the Story That Inspired 'The Exorcist,'" *Strange Magazine*, no. 20, 1999.

John Palmer, "A Mail Survey of Ouija Board Users in North America," *International Journal of Parapsychology*, vol. 12, no. 2, 2001.

Monica Davey, "Faith Healing Gone Wrong Claims Boy's Life," *New York Times*, August 29, 2003.

Amy C. Clark, "The Legend of Bouncing Bertha," *Blue Ridge Country*, 2004.

Crucified Romanian nun exhumed, BBC News, September 21, 2005.

Barbie Nadeau, "The Devil in Pictures," *Newsweek International*, October 24, 2005.

Andrew B. Newberg, Nancy A. Wintering, Donna Moran, Mark R. Waldman, "The Measurement of Regional Cerebral Blood Flow During Glossolalia: A Preliminary SPECT Study," *Psychiatric Research: Neuroimaging*, vol. 148, no. 1, November 2006.

Letters and materials from the Parapsychology Laboratory Records, 1893–1984, Rare Book, Manuscript, and Special Collections Library, Duke University, Durham, North Carolina:

Rev. Luther M. Schulze to J. B. Rhine, March 21, 1949.

Mrs. J. B. Rhine to Rev. Luther M. Schulze, March 23, 1949.

J. B. Rhine to Rev. Luther M. Schulze, April 2, 1949.

Rev. Luther M. Schulze to J. B. Rhine, April 6, 1949.

Rev. Luther M. Schulze to J. B. Rhine, April 19, 1949.

J. B. Rhine to Rev. Luther M. Schulze, April 21, 1949.

J. B. Rhine to Rev. Luther M. Schulze, April 25, 1949.

J. B. Rhine to Richard Darnell, May 10, 1949.

J. B. Rhine to Rev. Edward Dahmus, July 7, 1949.

J. B. Rhine to Richard Darnell, July 8, 1949.

J. B. Rhine to Richard Darnell, August 17, 1949.

J. B. Rhine to Rev. Luther M. Schulze, September 15, 1949.

Rev. Luther M. Schulze to J. B. Rhine, September 19, 1949.

"Dianetics is not based on any established scientific work so far as I know." From J. B. Rhine letter dated November 7, 1950.

John Freeman answers a request about Ouija boards, November 15, 1960.

CHAPTER SIX

Interviews with or information supplied by Marjorie Bayless, Tom Butler, Garrett Husveth, Raoul J. Larrinaga, Ann Longmore-Etheridge, Elizabeth McMahan, Alexander MacRae, retired

Deputy Chief Kevin J. Mullen, Paolo Presi, Sally Rhine Feather, William Roll, Gertrude Schmeidler, Klaus Schmidt-Koenig, Robert Van de Castle, Rhea White, Pam Zimmer, Nancy L. Zingrone.

Morey Bernstein, *In Search of Bridey Murphy: With new material by William J. Barker*, Doubleday, 1965.

Eileen Coly, *Eileen J. Garrett: Adventures in the Supernormal*, Helix Press, 2002.

Sarah Estep, *Voices of Eternity*, Ballantine Books, 1988.

Alan Gauld, *Mediumship and Survival: A Century of Investigations*, David & Charles, 1983.

Naomi A. Hintze and J. Gaither Pratt, *The Psychic Realm: What Can You Believe?* Random House, 1975.

Hans Holzer, *The Psychic World of Bishop Pike*, Crown Publishers, 1970.

James A. Pike, *The Other Side: An Account of My Experiences with Psychic Phenomena*, Doubleday, 1968.

Vance Randolph, *Ozark Superstitions*, Columbia University Press, 1947.

Case Studies in Parapsychology in Honor of Dr. Louisa E. Rhine, ed. K. Ramakrishna Rao, McFarland & Company, 1986.

Louisa E. Rhine, *The Invisible Picture*, McFarland & Company, 1981.

———, *Something Hidden*, McFarland & Company, 1983.

David M. Robertson, *A Passionate Pilgrim*, Alfred A. Knopf, 2004.

D. Scott Rogo and Raymond Bayless, *Phone Calls from the Dead*, Berkley Publishing, 1980.

Carl Sagan, *The Demon-Haunted World: Science as a Candle in the Dark*, Ballantine Books, 1997.

Daniel B. Smith, *Muses, Madmen, and Prophets: Rethinking the History, Science, and Meaning of Auditory Hallucination*, Penguin Press, 2007.

"The Family's Last Word," *New York Times*, May 24, 1934.

"Body Will Be Exhumed," *New York Times*, May 25, 1934.

"Scientists Agree Man's 'Hunches' May Be Only '6th Sense' at Work," *New York Times*, December 31, 1949.

Aldous Huxley, "A Case for ESP, PK and PSI," *Life*, January 11, 1954.

Alfred Hitchcock, "My Five Greatest Mysteries," *Coronet*, September 1955.

"Dean Pike Deplores Kind of Immortality Pictured by 'Bridey,'" *New York Times*, April 23, 1956.

Ashley Montagu, "Expert Opinion," an editorial about *A Scientific Report on "The Search for Bridey Murphy,"* edited by Milton V. Kline, *New York Times*, June 24, 1956.

Eileen Garrett, "The Bridge of Emotion," *Tomorrow*, vol. 4, no. 2, Winter 1956.

Louisa E. Rhine, "Hallucinatory PSI Experiences II. The Initiative of the Percipient in Hallucinations of the Living, Dying, and the Dead," *Journal of Parapsychology*, vol. 21, no. 1, March 1957.

Hornell Hart, "Do Apparitions of the Dead Imply Any Intention on the Part of the Agent? A Rejoinder to Louisa E. Rhine," *Journal of Parapsychology*, vol. 22, no.1, March 1958.

Raymond Bayless's letter to the editor, *Journal of the American Society for Psychical Research*, vol. 53, 1959.

Louisa E. Rhine, "Auditory PSI Experiences: Hallucinatory or Physical?" *Journal of Parapsychology*, vol. 27, no, 3, September 1963.

Bob Loftin, *The Tri-State Spook Light Authentic Guide*, self-published, 1963.

Raymond Bayless, "The Ozark Spook Light Article," *Fate*, September/October 1964.

W. D. Rees, "The Hallucinations of Widowhood," *British Medical Journal*, 1971.

"Nothing Hidden," *Time*, June 28, 1976.

D. J. Ellis, *The Mediumship of the Tape Recorder*, The Society for Psychical Research, 1978.

T. B. Posey and M. Losch, "Auditory Hallucinations of Hearing Voices in 375 Normal Subjects," *Imagination, Cognition, and Personality*, vol. 3, 1983.

P. R. Olson, J. A. Suddeth, P. J. Peterson, C. Egelhoff, "Hallucinations of Widowhood," *Journal of the American Geriatric Society*, August 1985.

T. R. Barrett and J. B. Etheridge, "Verbal Hallucinations in Normals, I: People Who Hear Voices," *Applied Cognitive Psychology*, vol. 6, no. 5, September/October1992.

Agneta Grimby, "Hallucinations Following the Loss of a Spouse: Common and Normal Events Among the Elderly," *Journal of Clinical Geropsychology*, vol. 4, 1998.

Mark Poysden and Ann Longmore-Etheridge, *EVP—Voices of the Dead*, 2000, retrieved from http://www.strangenation.com.au/Articles/sna_evp_vod.htm.

Alexander MacRae, "Report of an Anomalous Speech Products

Experiment Inside a Double Screened Room," *Journal of the Society for Psychical Research*, vol. 69.4, no. 881, October 2005.

Jennifer B. Ritsher, A. Lucksted, P. G. Otilingam, M. Grajales, "Hearing Voices: Explanations and Implications," *University of California Postprints*, Paper 1597, 2004.

Bruce Greyson, M.D., and Mitchell B. Liester, M.D., "Auditory Hallucinations Following Near-Death Experiences," *Journal of Humanistic Psychology*, vol. 44, 2004.

Carlos S. Alvarado, "Raymond Gordon Bayless Obituary," *Journal of the Society for Psychical Research*, vol. 69.4, no. 881, October 2005.

Ian Stevenson, "Half a Career with the Paranormal," *Journal of Scientific Exploration*, vol. 20, no. 1, 2006.

James E. Alcock, "Electronic Voice Phenomena: Voices of the Dead?" undated, CSI Web site, http://www.csicop.org/specialarticles/evp.html#author.

Letters and materials from the Parapsychology Laboratory Records, 1893–1984, Rare Book, Manuscript, and Special Collections Library, Duke University, Durham, North Carolina:

Gerald Heard to J. B. Rhine, April 24, 1952.

Bill Wilson to J. B. Rhine, November 6, 1952.

J. B. Rhine to Upton Sinclair, November 8, 1952.

J. B. Rhine to Dr. Beaumont S. Cornell, November 17, 1952.

Jackie Gleason to J. B. Rhine, October 16, 1953.

Eileen Garrett to J. B. Rhine, February 20, 1954.

J. B. Rhine to W. H. Belk Jr., March 15, 1954.

Morey Bernstein to W. H. Belk Jr., March 22, 1954.

J. B. Rhine to W. H. Belk Jr., March 30, 1954.

J. B. Rhine to Eileen Garrett, July 2, 1954.

J. B. Rhine to Adam Linzmayer, July 26, 1954.

Morey Bernstein to J. B. Rhine, September 13, 1954.

Eileen Garrett to J. B. Rhine, April 26, 1955.

J. B. Rhine to Mr. Henry Belk (a different Henry Belk), March 20, 1956.

Pearl S. Buck to J. B. Rhine, May 2, 1959.

Alfred P. Sloan Jr. to J. B. Rhine, August 10, 1959.

J. B. Rhine to Dr. Carroll B. Nash, November 27, 1961.

Raymond Bayless to J. B. Rhine, April 19, 1964.

J. B. Rhine to Raymond Bayless, May 8, 1964.

Everett F. Dagle to Morey Bernstein, October 16, 1963.

CHAPTER SEVEN

Interviews with or information supplied by Anna Barbay, Lisette
 Coly, Tom Edwards, Andrea Herrmann, Mary Jean Herrmann,
 Lucille Herrmann Patricia, David Kahn, Stanley Krippner, Carol
 Labate, Sally Rhine Feather, William Roll, Carla Tozzi, Robert
 Van de Castle, Rhea White, Officer Douglas Stiegelmaier (Nassau
 County Police Department George F. Maher Museum).

Detective Tozzi's case files from the Nassau County Police Museum.

Shirley Jackson, *The Haunting of Hill House*, Viking Penguin, 1959.

Jurgen Keil, ed., *Gaither Pratt: A Life for Parapsychology*, McFarland
 & Company, 1987.

J. Gaither Pratt, *Parapsychology: An Insider's View of ESP*, Doubleday
 & Company, Inc., 1964.

Louisa E. Rhine, *Something Hidden*, McFarland & Company, 1983.

William G. Roll, *The Poltergeist*, Paraview Special Editions, 2004.

Meyer Berger, "Quest for Haunted House Here Finds Ghosts Shun
 Metropolis of Steel and Concrete," *New York Times*, February 29,
 1956.

"Family Leaves Spooked House," *Dallas Times Herald*, August 3, 1957.

"Some Recent Parapsychology Cases," *Parapsychology Bulletin*, no. 32,
 November 1957.

Dave Kahn, "Flying-Bottle Mystery Gets Expert Attention,"
 Newsday, February 26, 1958.

Robert Wallace, "House of Flying Objects," *Life*, March 17, 1958.

"The Seaford Case," Parapsychology Foundation newsletter, vol. 5,
 no. 2, March–April 1958.

Karlis Osis, "An Evaluation of the Seaford Poltergeist Case,"
 Parapsychology Foundation newsletter, vol. 5, no. 2, March–April
 1958.

"Seaford Revisited: Post-Mortem on a Poltergeist, An Editorial
 Report," *Tomorrow*, vol. 6, no. 3, Summer 1958.

"L.I. Poltergeist Stumps Duke Men," *New York Times*, August 10,
 1958.

Dave Kahn, "Bottle-Popping Probe Winds Up with Fizzle,"
 Newsday, August 11, 1958.

W. G. Roll, "Some Physical and Psychological Aspects of a Series of
 Poltergeist Phenomena," *Journal of the American Society of Psychical
 Research*, vol. 62, no. 3, July 1968.

Dave Kahn, "A Home's Bad Vibration," *Newsday*, 2005.

Letters and materials from the Parapsychology Laboratory Records,

1893–1984, Rare Book, Manuscript, and Special Collections Library, Duke University, Durham, North Carolina:

J. B. Rhine to Francis Bolton, August 3, 1936.

Hans Holzer to J. B. Rhine, 1955.

J. B. Rhine to Hans Holzer, February 5, 1955.

J. B. Rhine to Eileen Garrett, February 12, 1955.

J. B. Rhine to W. P. Bentley, September 10, 1957.

J. B. Rhine to Sally and Ben Feather, November 3, 1958.

Eileen Garrett to Gaither Pratt, October 22, 1958.

Asst. Prof. Philip J. Lorenz to Gaither Pratt, December 31, 1960.

CHAPTER EIGHT

Interviews with or information supplied by Detective Andy Arostegui (Miami Police Department Cold Case Squad), Eileen Coly, Lisette Coly, Weston DeWalt, Detective Vivian Flores, Daryl F. Glaser (Counsel, Commonwealth of Massachusetts Department of Correction Legal Division), Stephanie Hurkos, Mrs. Joseph Kremen, Steven Levy, Joseph W. McMoneagle, Walter F. Rowe, Ph.D., (Department of Forensic Sciences, George Washington University), (Wisconsin Federal Court, Eastern District), Irving Whitman.

Norma Lee Browning, *The Psychic World of Peter Hurkos*, Doubleday, 1970.

———, *Peter Hurkos: I Have Many Lives*, Doubleday, 1976.

Edward Keyes, *Michigan Murders*, Pocket, reissue edition, 1990.

Steven Levy, *The Unicorn's Secret: Murder in the Age of Aquarius*, Prentice-Hall, 1988.

Arthur Lyons and Marcello Truzzi, *The Blue Sense: Psychic Detectives and Crime*, Mysterious Press, 1991.

Walter J. McGraw, "Where Hurkos Failed," in *The Satan Trap: Dangers of the Occult*, ed. Martin Ebon, Doubleday, 1976.

Mike Marinacci, *Mysterious California: Strange Places and Eerie Phenomena in the Golden State*, Panpipes Press, 1988.

Gordon Stein, ed., *Encyclopedia of the Paranormal*, Prometheus Books, 1996.

Lawrence Thompson, "Mentalist and Police Visit Death Scene," *Miami Herald*, May 25, 1957.

Kit Miniclier, "Telepathist Predicting Jackson Case Solution," Northern Virginia *Sun*, June 8, 1960.

"Herkos [*sic*] Gift 'God Given' Is Claim," Northern Virginia *Sun*, June 8, 1960.

"Jackson Slaying Investigation, Aided by Telepathist, Puts Man in Asylum," Washington, DC, *The Evening Star*, June 10, 1960.

Jeffrey S. O'Neil, "Telepathist's Suspect Faces Test in Murder," *Washington Post*, June 11, 1960.

Jeffrey S. O'Neil, "'Mindreader' in Jackson Case Quits Without Finding Any Evidence," *Washington Post*, June 12, 1960.

"Deputies Hunt Missing Boy, 7," *Los Angeles Times*, July 15, 1960.

"Bloody Cloth Found in Missing Boy Search," *Los Angeles Times*, July 16, 1960.

"5-Day Search Fails to Find Missing Boy," *Los Angeles Times*, July 18, 1960.

"Mountaineer Unit in Search for Boy," *Los Angeles Times*, July 19, 1960.

S. H. Posinsky, Ph.D., "The Case of John Tarmon: Telepathy and the Law," *Psychiatric Quarterly*, vol. 3, no. 1, March 1961.

Ben A. Franklin, "Psychist Failed in Virginia Case," *New York Times*, February 9, 1964.

"Mental 'Expert' Is Indicted on Impersonation Charge," *New York Times*, March 4, 1964.

"Psychist Convicted of Impersonating an Agent of F.B.I," *New York Times*, November 14, 1964.

Henry K. Puharich, "Electrical Field Reinforcement of ESP," *International Journal of Neuropsychiatry*, vol. 2, no. 5, October 1966.

Dave Smith, "Suspect in Child Killings Called 'Quiet, Nice Guy' by Neighbors," *Los Angeles Times*, March 8, 1970.

Tom Newton, "Body Believed That of Girl, 8, Slain 17 Years Ago Found," *Los Angeles Times*, March 12, 1970.

Jim Stingley, "Slayer of Six Children Hangs Himself in Cell," *Los Angeles Times*, October 31, 1971.

J. A. Sweat and M. W. Durm, "Psychics: Do Police Departments Really Use Them?" *Skeptical Inquirer*, Winter 1993.

J. Nickell, "Update: Psychics—Do Police Departments Really Use Them in Small and Medium-sized Cities?" in *Psychic Sleuths*, ed. by M. W. Durm and J. A. Sweat, Prometheus Books, 1994.

Marcello Truzzi, "Reflections on 'The Blue Sense' and Its Critics," *Journal of Parapsychology*, vol. 59, no. 2, June 1995.

Jill Neimark, "Do the Spirits Move You?" *Psychology Today*, September/October 1996.

Andrew Blankstein, "Killer's Dead, but They're Still on His Trail," *Los Angeles Times*, March 17, 2007.

Kenneth Todd Ruiz, "Police Back Theory on Missing Boy," Whittier (CA) *Daily News*, March 19, 2007.

Malcolm Gladwell, "Dangerous Minds," *New Yorker*, November 12, 2007.

Letters and materials from the Parapsychology Laboratory Records, 1893–1984, Rare Book, Manuscript, and Special Collections Library, Duke University, Durham, North Carolina:

Mary Craig Sinclair to J. B. Rhine, September 6, 1945.

Joseph Kremen to J. B. Rhine, November 1960.

J. B. Rhine to Joseph Kremen, November 15, 1960.

Harold Sherman to J. B. Rhine, November 22, 1960.

Joseph Kremen to J. B. Rhine, January 1961.

J. B. Rhine to Joseph Kremen, January 27, 1961.

S. H. Posinsky to J. B. Rhine, February 15, 1961.

M. G. Beard to J. B. Rhine, February 27, 1961.

Harold Sherman to J. B. Rhine, March 1, 1961.

Gus Turbeville to J. B. Rhine, November 28, 1961.

Gus Turbeville to Steven Rockefeller, November 28, 1961.

Morey Bernstein to J. B. Rhine, April 17, 1962.

J. B. Rhine to Sollie H. Posinsky, June 7, 1962.

Morey Bernstein to J. B. Rhine, March 4, 1963.

J. B. Rhine to Morey Bernstein, July 10, 1963.

Gus Turbeville to Morey Bernstein, September 19, 1963.

Morey Bernstein to J. B. Rhine, September 27, 1963.

W. H. Belk memo dated October 5, 1963, Subject: Peter Hurkos today.

Kay Sterner to Dr. Andrija Puharich, September 2, 1964.

Katherine Ramsland, "John Norman Collins: The Co-ed Killer," retrieved from http://trutv.com/library/crime/serial_killers/predators/collins/body_1.html.

http://www.crimelibrary.com/criminal_mind/profiling/steven_egger/2.html

http://www.crimelibrary.com/criminal_mind/forensics/psychics/9.html

www.peterhurkos.com

CHAPTER NINE

Interviews with or information supplied by Eileen Coly, Lisette Coly, Stanley Krippner, Richard Lowrie, Joseph W. McMoneagle, Sally Rhine Feather.

Jurgen Keil, ed., *Gaither Pratt: A Life for Parapsychology*, McFarland & Company, 1987.

Martin A. Lee and Bruce Shlain, *Psychedelic Pioneers*, Grove Weidenfeld, 1985.

J. Gaither Pratt, *Parapsychology: An Insider's View of ESP*, Doubleday, 1964.

Jon Ronson, *The Men Who Stare at Goats*, Simon & Schuster, 2004.

Aldous Huxley, "Miracle in Lebanon," *Esquire*, September 1955.

J. B. Rhine, "Why National Defense Overlooks Parapsychology," *Journal of Parapsychology*, vol. 21, no. 4, December 1957.

Henry K. Puharich, "Can Telepathy Penetrate the Iron Curtain," *Tomorrow*, vol. 5, no. 2, Winter 1957.

Bess Furman, "Scientists Urge Behavior Study," *New York Times*, February 9, 1958.

Arthur Koestler, "Return Trip to Nirvana," (London) *Sunday Telegraph*, March 12, 1961.

Humphrey Osmond, "Peyote Night," *Tomorrow*, vol. 9, no. 2, Spring 1961.

Sidney Katz, "ESP: First Report on Extrasensory Powers among Canadians," *Maclean's*, July 29, 1961.

Karlis Osis, "Deathbed Observations," Physicians and Nurses, Parapsychology Foundation, 1961.

————, "A Pharmacological Approach to Parapsychological Experimentation," *Proceedings of Two Conferences on Parapsychology and Pharmacology*, Parapsychology Foundation, 1961.

Eileen Garrett, "Psychopharmacological Parallels to Mediumship," *Proceedings of Two Conferences on Parapsychology and Pharmacology*, Parapsychology Foundation, 1961.

Research Bulletin, Air Force Cambridge Research Laboratories, September 1962.

"Parapsychology in Russia and Czechoslovakia," *Journal of the Society for Psychical Research*, vol. 42, no. 715, March 1963.

William R. Smith, Everett F. Dagle, Margaret D. Hill, John Mott-Smith, "Research Report: Testing for Extrasensory Perception with a Machine," Air Force Cambridge Research Laboratories, May 1963.

"U.S. Defense Has ESP Machine," *Parapsychology Bulletin*, no. 6, August 1963.

Eugene B. Konecci, Ph.D., "Bioastronautics Review," a paper presented at the XIV International Astronautics Federation Meeting in Paris, September 29–October 1, 1963.

Duncan Blewett, "Psychedelic Drugs in Parapsychological Research," *International Journal of Parapsychology*, vol. 5, no. 1, Winter 1963.

"Toward Telepathy in Outer Space," *Parapsychology Bulletin*, January/February 1964.

J. Gaither Pratt, "Extrasensory Perception in Russia and Czechoslovakia," *International Journal of Neuropsychiatry*, September/October 1966.

J. M. Utts, Response to Ray Hyman's Report of September 11, 1995, "Evaluation of Program on Anomalous Mental Phenomena," 1996, http://anson.ucdavis.edu/%7Eutts/response.html.

Brookings Institution, U.S. Nuclear Weapons Cost Study Project, 1998.

Arthur Hastings, "The Many Voices of Eileen J. Garrett," *International Journal of Parapsychology*, vol. 12, no. 2, 2001.

"Monkey Ward," a confidential CIA memo dated April 20, 1954, and released January 21, 2002. Author's name blacked out.

David R. McLean, "Cranks, Nuts, and Screwballs," *Studies in Intelligence*, CIA, Vol. 9, Summer 1965.

Letters and materials from the Parapsychology Laboratory Records, 1893–1984, Rare Book, Manuscript, and Special Collections Library, Duke University, Durham, North Carolina:

Reynold B. Johnson to J. B. Rhine, April 5, 1938.

Andrija Puharich to J. B. Rhine, June 1, 1953.

J. B. Rhine to Aldous Huxley, August 15, 1957.

Aldous Huxley to J. B. Rhine, September 19, 1957.

Peter A. Castruccio to J. B. Rhine, November 15, 1959.

Col. William H. Bowers to J. B. Rhine, July 14, 1958.

P. A. Castruccio to J. B. Rhine, August 19, 1958.

Peter A. Castruccio to J. B. Rhine, November 15, 1959.

Col. William H. Bowers to J. B. Rhine, June 6, 1960.

Barbara B. Brown to J. B. Rhine, February 16, 1961.

J. B. Rhine to B. W. Russell, Wright Air Development Division, February 24, 1961.

Lt. Col. George J. Bayerle Jr. to J. B. Rhine, February 28, 1961.

George H. Scheer to J. B. Rhine, March 17, 1961.

Edwin G. Boring to J. B. Rhine, March 25, 1961.

J. B. Rhine to Arthur Koestler, May 10, 1961.

Arthur Koestler to J. B. Rhine, May 26, 1961.

Chester Carlson to J. B. Rhine, May 29, 1961.

J. B. Rhine to Timothy Leary, June 5, 1961.

Timothy Leary to J. B. Rhine, June 5, 1961.

J. B. Rhine to Timothy Leary, June 9, 1961.

Chester Carlson to J. B. Rhine, June 15, 1961.

J. B. Rhine to Timothy Leary, June 20, 1961.

Timothy Leary to J. B. Rhine, June 28, 1961.

John Altrocchi to Timothy Leary, July 5, 1961.

Timothy Leary to J. B. Rhine, October 5, 1961.

J. B. Rhine to Timothy Leary, November 25, 1961.

J. B. Rhine to Chester Carlson, December 4, 1961.

Hans F. Meissinger to J. B. Rhine, December 26, 1961.

Stanley Krippner to J. B. Rhine, March 30, 1962.

J. B. Rhine to Mr. Stout (*Chicago Daily News*), May 17, 1962.

Gaither Pratt address to everyone at the lab, June 8, 1962.

Ivan Tors to J. B. Rhine, June 22, 1962.

J. B. Rhine to W. P. Bentley, June 23, 1962.

J. B. Rhine to Dr. Jesse Orlansky, November 26, 1962.

Everett F. Dagle to J. B. Rhine, 1963 (not specifically dated).

A. L. Kitselman to J. B. Rhine, September 16, 1963.

J. B. Rhine to Dr. E. B. Konecci, December 10, 1963.

Richard Trumbull, Director Psychological Sciences Division, Department of the Navy, Office of Naval Research, to J. B. Rhine, October 20, 1964.

Eileen Garrett to Aldous Huxley, January 14, 1959, from the Eileen J. Garrett Library, Parapsychology Foundation Library, Greenport, NY.

CHAPTER TEN

Interviews with or information supplied by Carlos S. Alvarado, Lloyd Auerbach, Eileen Coly, Lisette Coly, Brenda J. Dunne, Carole Linda Gonzalez, Hans Holzer, Robert G. Jahn, Patricia McMaster, Dr. Michael Persinger, Sally Rhine Feather, William Roll, Wendy Viernow, Nancy L. Zingrone.

Lloyd Auerbach, *A Paranormal Casebook: Ghost Hunting in the New Millennium*, Atriad Press, 2005.

C. J. Ducasse, *A Critical Examination of the Belief in a Life After Death*, Charles C. Thomas, 1961.

Chris Carter, *Parapsychology and the Skeptics*, Sterling House Publishers, 2007.

Alan Gauld, *Mediumship and Survival: A Century of Investigations*, David & Charles, 1983.

Constance M. Greiff, *The Morris-Jumel Mansion: A Documentary History*, Heritage Studies (undated).

Marianne Hancock, *Madame of the Heights*, Windswept House, 1998.

Hans Holzer, *Ghosts: True Encounters with the World Beyond*, Black Dog & Leventhal Publishers, 2004.

Dean Radin, *Entangled Minds*, Pocket Paraview, 2006.

Bruce Rosenblum and Fred Kuttner, *Quantum Enigma: Physics Encounters Consciousness*, Oxford University Press, 2006.

Michael Schermer, *Why People Believe Weird Things*, A. W. H. Freeman/Owl Book, 2002.

William Henry Shelton, *The Jumel Mansion*, Houghton Mifflin, 1916.

The Jumel Papers at the New York Public Library and the New-York Historical Society.

The Morris-Jumel Archives at the Morris-Jumel Mansion.

Eliza Jumel obituary, *New York Times*, July 18, 1865.

"Rinn Offer Rubbish," *New York Times*, January 29, 1920.

William Henry Shelton, "New Clues in the Enigma of Mme. Jumel's Life," *New York Times*, May 13, 1928.

"Ex Boxer Slain," Newark *Evening News*, December 14, 1956.

"Mrs. Ernest Rivers Held in Gun Death," Newark *Evening News*, December 15, 1956.

"Ghost Walks Again," Newark *Evening News*, May 13, 1961.

Eileen Garrett, "The Nature of My Controls," *Tomorrow*, vol. 11, no. 4, Autumn 1963.

Harry Altshuler, "Ghost Hunt Flops on Napoleon's Bed," *New York World-Telegram*, January 24, 1964.

Grace Glueck, "2 Uptown 'Ghosts' Get Eviction Call," *New York Times*, May 23, 1964.

William P. Schweickert Jr., "A Seance at Jumel Mansion," *Westchester Historian*, vol. 41, no. 2, Spring 1965.

W. G. Roll, "The Newark Disturbances," *Journal of the American Society for Psychical Research*, vol. 63, no. 2, April 1969.

Richard A. Kalish and David K. Reynolds, "Phenomenological Reality and Post-Death Contact," *Journal for the Scientific Study of Religion*, vol. 12, no. 2, June 1973.

Anna Quindlen, "About New York: Belief If Ghost Haunts a Historic Mansion," *New York Times*, October 31, 1981.

Victor J. Stenger, "Quantum Quackery," *Skeptical Inquirer*, January 1997.

Michael A. Persinger, "The Neuropsychiatry of Paranormal Experiences," *Journal of Neuropsychiatry and Clinical Neurosciences*, vol. 13, no. 4, Fall 2001.

Arran Frood, "Ghostly Magnetism Explained," BBC News, May 21, 2003.

L. S. St. Pierre and M. A. Persinger, "Experimental Facilitation of the Sensed Presence Is Predicted by the Specific Patterns of the Applied Magnetic Fields, Not by Suggestibility: Re-Analyses of 19 Experiments," *International Journal of Neuroscience*, vol. 116, 2006, pp. 1079–96.

William G. Roll and William T. Joines, "Energetic Aspects of RSPK," paper presented at the Parapsychological Association Convention, 2007.

Robert Lanaza, "A New Theory of the Universe," *American Scholar*, Spring 2007.

Robert G. Jahn and Brenda J. Dunne, "Change the Rules," *Journal of Scientific Exploration*, vol. 22, no. 2, Spring 2008.

Letters and materials from the Parapsychology Laboratory Records, 1893–1984, Rare Book, Manuscript, and Special Collections Library, Duke University, Durham, North Carolina:

J. B. Rhine to Dr. Albert Einstein, July 11, 1940.

Albert Einstein to J. B. Rhine, July 23, 1940.

J. B. Rhine to Professor M. A. Wenger, December 21, 1961.

J. B. Rhine to Chester Carlson, March 16, 1964.

J. B. Rhine to James Randi, July 23, 1964.

CSICOP Release on Gallup Poll, retrieved from www.csicop.org, June 11, 2001.

CHAPTER ELEVEN

Interviews with or information supplied by Jim Carpenter, Elizabeth McMahan, Joanna Morris, Sally Rhine Feather, and material from Jim Carpenter's February 7, 1983, interview with Dorothy Pope.

Allan Agoff, *Eileen Garrett and the World Beyond the Senses*, Harper and Row, 1974.

Anthony Arthur, *Radical Innocent*, Random House, 2006.

Chris Carter, *Parapsychology and the Skeptics*, Sterling House Publishers, Inc., 2007.

William James, *The Will to Believe*, Longman, Green & Co., 1897.

Elizabeth McMahan, *Warming Both Hands Before the Fire of Life*, vol. 3, self-published, 2005.

J. Gaither Pratt, *Parapsychology: An Insider's View of ESP*, Doubleday, 1964.

Ira Progoff, *The Image of an Oracle: A Report on Research into the Mediumship of Eileen J. Garrett*, Garrett Publications, 1964.

J. B. Rhine and Associates, *Parapsychology: From Duke to FRNM*, The Parapsychology Press, 1965.

Louisa E. Rhine, *Something Hidden*, McFarland & Company, 1983.

J. B. Rhine, Ph.D., and Louisa E. Rhine, Ph.D., "One Evening's Observation on the Margery Mediumship," *Journal of Abnormal and Social Psychology*, vol. 21, no. 4, January–March 1927.

Milton Esterow, "E.S.P. Mail Floods 'Clear Day,'" *New York Times*, January 27, 1966.

Persi Diaconis, "Statistical Problems in ESP Research," *Science*, vol. 201, 1978.

Report of the Executive Director to the Board of Directors of the Foundation for the Research on the Nature of Man, April 24, 1965.

Marcello Truzzi, "Discussion on the Reception of Unconventional Scientific Claims," in *AAAS Symposium, The Reception of Unconventional Science*, ed. Seymour H. Mauskopf, Westview Press, 1979.

Marcello Truzzi, "On Pseudo-Skepticism: A Commentary," *Zetetic Scholar*, no. 12–13, 1987.

E. Douglas Dean, Ph.D., "20th Anniversary of the PA and the AAAS, Part 1: 1963–1969," *ASPR Newsletter*, Winter 1990.

John Archibald Wheeler, "Drive the Pseudos out of the Workshop of Science," Appendix A in *The Role of Consciousness in the Physical World*, ed. by Robert G. Jahn, Westview Press, 1981.

Minutes from the Great Issues of Conscience in Modern Medicine conference at Dartmouth, September 8–10, 1960, Dartmouth College Library, Rauner Special Collections Library.

Letters and materials from the Parapsychology Laboratory Records, 1893–1984, Rare Book, Manuscript, and Special Collections Library, Duke University, Durham, North Carolina:

Gaither Pratt to Hubert Pearce, May 28, 1964.

J. B. Rhine to Clement Stone, July 27, 1965.

J. B. Rhine to Clement Stone, August 2, 1965.

J. B. Rhine to Terry Sanford, December 1, 1978.

J. B. Rhine to Gaither Pratt, August 21, 1979.

Vernon Pratt letter to the editor, *The Chronicle*, Durham, NC, December 16, 1996.

EPILOGUE

Donald K. Adams, "The Natural History of a Prejudice," Archives for the History of American Psychology, University of Akron.

Persi Diaconis, "Statistical Problems in ESP Research," *Science*, New Series, vol. 201, no. 4351, July 14, 1978.

Andrei Linde, "Inflation, Quantum Cosmology, and the Anthropic Principle," *Science and Ultimate Reality: From Quantum to Cosmos*, honoring John Wheeler's 90th birthday, Cambridge University Press, 2003.

ACKNOWLEDGMENTS

This book would have been impossible to write without the help of my agent, Betsey Lerner; my editor, Lee Boudreaux, and her incredible assistant, Abigail Holstein; my friend Howard Mittelmark; Sally Rhine Feather, Rosie Rhine, Rob Rhine, Eileen Coly, Lisette Coly, and Ellen Pratt, the children (and one grandchild) of my subjects; the former Parapsychology Laboratory researchers Betty McMahan, the late Rhea White, William Roll, and Robert Van de Castle; the parapsychologists Jim Carpenter, Nancy L. Zingrone, and Carlos S. Alvarado; the former PEAR Laboratory Director Robert G. Jahn and PEAR Laboratory Manager Brenda J. Dunne; and the indispensable assistance of the librarians at the Special Collections Library of Duke University: Eleanor Mills, Elizabeth Dunn, Janie Morris, Zach Elder, and Linda McCurdy.

INDEX

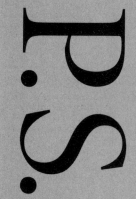

Insights,
Interviews
& More...

Meet Stacy Horn

I'D LIKE TO BEGIN WITH A GHOST STORY.

When I was six or seven, I had wandered off from home and lost my way. I wasn't afraid. I was always drifting off to explore and getting lost, but I always managed to find my way home. This time I passed by an elderly woman in her yard with a bunch of cats. I love cats and I asked if it was all right if I pet them. She said of course, and after a few minutes she very kindly invited me inside for milk and cookies. When I was done she walked me home. It turned out I was only a few blocks away from my house. My mother punished me for straying too far, and that was that.

A year later I decided to go back and visit the lady and her cats. But when I got to the house the whole place was in complete disrepair. The roof was partially caved-in and the front yard was wild and overgrown. I stood there trying to understand what I was seeing. I knew even at that young age that this was a lot of damage and that the place had been neglected for a long time. I also knew this meant that something was weird about my visit with the lady. So I just stood there, completely flummoxed. After a few minutes a woman came out of the house next door and asked me if she could help me.

"Where is the lady?"

"What lady?"

"The lady who lives here."

Then she said very gently, "No one has lived here for a very long time."

I wasn't surprised when she told me that, but I was still very confused. I told her about the old lady and the cats, and she told me that the old lady hadn't lived there for years. I knew she was telling me that the lady was dead and that she was dead before I had met her. Again I wasn't scared. I don't think I really understood death yet. I also wasn't thinking ghosts. For me ghosts were cartoons, like Casper, and the lady I met was a person, a solid, breathing, talking person. With cats. I had seen her, talked to her, played with her cats, sat at her table, and ate her cookies, and when she walked me home I had held her hand. There was nothing spooky about my encounter; it was all completely ordinary. I couldn't explain it, and so I continued to stand there, trying to make sense of it all. I didn't think she was a ghost, but then there had to be another explanation.

I told the next-door neighbor that I was fine, and when she left I went inside the old woman's house to investigate. Maybe she was hiding. But the place was a wreck. The table at which I had sat with my milk and cookies was still there, but it was piled with debris that had come in from a hole in the roof. The whole interior was well on its way to being taken over by the elements. Everything I saw seemed to confirm that I couldn't have met anyone here. This place had been abandoned for a long time. But I was still only perplexed, not frightened. Then I heard a voice. It was very close, like someone was talking quietly right beside me. It said, "Look in the bedroom." Or "Go into the bedroom." You would think when one hears a disembodied voice she'd remember the exact wording, but, once again, I was not scared or even particularly startled. I went into the bedroom. ▶

> ❝ For me ghosts were cartoons, like Casper, and the lady I met was a person, a solid, breathing, talking person. With cats. . . . I didn't think she was a ghost, but then there had to be another explanation. ❞

On the floor was a brown carton, like the kind you get from a supermarket. And inside the cartoon were four kittens. The kittens were a present from the old lady, I decided. I can't say I "knew" that, it was just the explanation I gave myself for the voice and the presence of a box of kittens. It was also an explanation that served my immediate purposes: I wanted those kittens. I picked up the box, took the kittens home, asked my mother if I could keep them, and she amazingly said yes.

Looking back, I think I wasn't frightened because as a kid, all the world was weird and unexplainable and I pretty much took everything in indiscriminately, trying to learn. I also think that because this experience had such a happy outcome it made me open to ghosts and anything paranormal. What's so scary about ghosts if they lead you to kittens?

When I was researching this book I couldn't help but hope that I'd find definitive, incontrovertible evidence for life after death. Did I really see a ghost that day? The answer is: maybe. The Duke experiments did not definitely answer that question, but I found a lot of stories that can't be explained and scientific results that still require enlightenment, all of which could be evidence for something that continues after we're gone—but there are other explanations. And for me, those explanations might be just as exciting.

Like professor Bernard Riess said all those years ago, when his colleagues criticized him for publishing positive results from his ESP tests, I think it's a mistake to throw away good data just because we don't yet understand their full implications. Suppressing the research hasn't been useful for either side of the debate. Had the scientists of Dr. Louisa Rhine's day paid attention to her papers about audio

> 66 Looking back, I think I wasn't frightened because as a kid, all the world was weird and unexplainable and I pretty much took everything in indiscriminately, trying to learn. 99

hallucinations, for instance, the recent "discoveries" being made in this area today—that people hear voices more often than we knew, and that this isn't necessarily a sign of mental illness—would have begun fifty years ago and we would be that much further along in understanding what is happening and why.

While I was pretty bleak at the end of the book about the current prospects for parapsychology, it would probably be more accurate to say that in many ways parapsychology has been at the forefront of consciousness research and that parapsychology and other areas of science such as neuroscience, quantum physics, and biology are coming together to a certain extent and finding a common ground there. Dr. Andrew Newberg recently scanned the brains of mediums in Brazil. The Institute of Noetic Sciences has ongoing projects looking at consciousness and the material world and healing; the Rhine Research Center is developing a theory and a program of research to explain PSI as an unconscious process that is always at work. I've noticed another promising development. With the exception of psychology, the emotions have always been off-limits for science. They were just too squishy and difficult to pin down and quantify. But now emotions are also all the rage and scientists from other disciplines are looking at them in a more serious way. The researchers at the Duke Parapsychology Laboratory always felt emotions were a crucial element of ESP.

It now seems possible to me that a future "grand unification theory" about consciousness and emotions will eventually incorporate the past and ongoing contributions of parapsychology. ∽

When a Good Story Won't Let You Go

WHEN YOU HAVE A GOOD STORY it's hard to let go. Since my book came out I have continued to research this topic and the stories I investigated. I follow the progress of the investigation into the murder of six-year-old Bruce Kremen (still ongoing). I keep on top of recent scientific developments that might explain the results of the Parapsychology Lab. For instance, evidence for something called mirror neurons has recently been found. These are neurons in your brain that activate in situations such as your seeing someone in distress. The mirror neurons mimic the neurons that are activating in the brain of the person you are watching. Perhaps information travels in this way, and this might one day explain how people sometimes know things they have no other way of knowing. Likewise, successful experiments in quantum entanglement have involved progressively larger particles at greater distances.

I've noticed that unlike in Rhine's time, careers aren't over when a scientist expresses an open mind about ESP. Physicist Michio Kaku wrote in his book *Physics of the Impossible* that telepathy doesn't violate the laws of physics as we know them and is therefore possible, although what he describes as telepathy is more of a mechanical telepathy, which is different from how the Duke scientists imagine it. But the Army is now funding research into this "synthetic telepathy" at the University of California, Irvine, and the Defense Advanced Research Projects Agency (DARPA) is pursuing another telepathic simulation called "silent talk."

Then there are the multiverse theories from physicists Hugh Everett and now Andrei Linde,

> **"** I've noticed that unlike in Rhine's time, careers aren't over when a scientist expresses an open mind about ESP. **"**

although they're talking about two entirely different things. Everett's theory, formulated in the fifties, proposes that there are many universes, which are endlessly branching off, creating new universes, each a more or less different version of the universe it branched off from. Linde, who has been refining his thinking since the seventies, theorizes that right after the big bang a phenomena called inflation created many universes, each with its own laws of physics, which don't necessarily look anything like our own and if they do it's purely accidental. Linde also proposes that this process of inflation is continuing, and somewhere out there universes continue to come about as a result of inflation. Can the presence of other universes explain the sources of information we appear to be able to sense or access?

I also especially love fact-checking séances from the past. We have better tools now for researching information, so we can go back and see how well the mediums did. My favorite is a séance that actually took place only a few blocks from where I live.

On June 26, 1957, the *New York Times* ran a piece by Meyer Berger called "Ghostly Coincidences Puzzle Bohemian Couple in 125-Year-Old House in Greenwich Village" about a haunted house in the West Village. Briefly: Harvey Slatin bought the red brick house and was in the process of converting it from a rooming house into a single-family home. The Slatins and the construction workers sometimes heard what sounded like a woman on the stairs. At first they just figured they had an intruder, and they'd wait to hear the sounds and then run upstairs. But no one was ever there. Their carpenter wrote it off to the odd sounds you hear in old homes. Slatin wasn't particularly unnerved either, and instead he tried to study the phenomena. He timed ▶

7

her ghostly steps and noted that they began at eleven in the morning and continued off and on until dusk. "I'd call them rather friendly sounds; a wee bit spooky, maybe," he said, "but somehow not frightening."

Later, when the carpenter was removing the ceiling on the top floor, a small tin about the size of a can of coffee fell onto his head. The label read, "The last remains of Elizabeth Bullock, deceased. Cremated January 21, 1931." Slatin called the crematorium listed on the container and learned that Elizabeth Bullock had been hit by a car on Hudson Street and taken to a drugstore nearby, where she died. She had lived on Perry Street though, and no one could explain how she ended up in a ceiling on Bank Street. Other than these few small facts, Slatin wasn't able to learn anything else about her.

Ghost hunter Hans Holzer (who died on April 26, 2009) read Berger's piece and contacted the Slatins to offer his services. During a séance conducted in the building, Holzer and his favorite medium, Ethel Johnson Meyers, came up with more (alleged) information about Elizabeth Bullock, which Holzer wrote about and published and which I have since researched. This particular séance had an interesting mix of hits and misses. One thing that came out of the séance was Elizabeth Bullock's wish to have a Christian burial, which she said she had been denied because she married outside her faith. Holzer's account ends with his suggestion that the Slatins bury Elizabeth in their garden. The Slatins say they're going to keep the tin containing Elizabeth's ashes displayed on the piano, where she's happy, they insist, but also in case someone shows up to claim her.

> 66 Later, when the carpenter was removing the ceiling on the top floor, a small tin about the size of a can of coffee fell onto his head. The label read, 'The last remains of Elizabeth Bullock, deceased. Cremated January 21, 1931.' 99

I tracked down Harvey Slatin in 2007, and he turned out to be Dr. Harvey Slatin, a Manhattan Project nuclear physicist. He was now ninety-two years old, and not terribly interested in talking to me, but he did tell me enough so that I could research the story more thoroughly. More important, although Dr. Slatin doesn't believe in ghosts he confirmed the unexplained events at Bank Street, whatever the final explanation may turn out to be.

The séance at Bank Street had taken place on a weekday evening in July. Mrs. Meyers went into a trance and immediately connected with a spirit named Betty, whom she said was paralyzed on one side and walked with a limp. Slatin's wife, Yeffe, was thrilled. She told them that she'd seen a lady with a limp with her "psychic eye." The spirit named Betty told her story, but like many stories told via mediums, her narrative is confusing. "He didn't want me in the family plot—my brother—I wasn't even married in their eyes. . . . But I was married before God . . . Edward Bullock . . . I want a Christian burial in the shades of the Cross . . . anyplace where the Cross is—but not with them!" Betty gave a few details about her life: her mother's maiden name was Elizabeth McCuller, and they came from Pleasantville, New York. When asked why her ashes were in the attic of Bank Street she gave an answer that didn't clarify anything. "I went with Eddie. There was a family fight . . . my husband went with Eddie . . . steal the ashes . . . pay for no burial . . . he came back and took them from Eddie . . . hide ashes . . . Charles knew it . . . made a roof over the house . . . ashes came through the roof . . . so Eddie can't find them."

It's all a bit impenetrable: "Just because I loved a man out of the faith, and so they ▶

> ❝ The séance at Bank Street had taken place on a weekday evening in July. . . . What's interesting is how much [medium] Ethel Johnson Meyers got right. ❞

When a Good Story Won't Let You Go
(continued)

took my bones and fought over them, and then they put them up in this place, and let them smolder up there, so nobody could touch them . . ." Who cremated you? Holzer asked. "It was Charles's wish, and it wasn't Eddie's, and therefore they quarreled. Charlie was a Presbyterian . . . and he would have put me in his church, but I could not offend them all. They put it beyond my reach through the roof; still hot . . . they stole it from the crematory."

A few more facts emerged. "Betty" had two children: Eddie, who was alive and living in California, and Gracie, who died as a baby. Also, "Betty" spoke the entire time with an Irish brogue. Holzer said he could tell an actor from the real thing, and this was the real thing. The spirit's last words were, "Lived close by. Bullock."

Hans Holzer and the Slatins didn't have the benefit of the Internet and resources such as Ancestry.com to help them research Elizabeth's history. Also, enough time has gone by that Elizabeth's death certificate is now in the public domain. What's interesting is how much Ethel Johnson Meyers got right. Elizabeth's husband was Edward Bullock. In Holzer's account he said her husband was Charlie, which was a reasonable guess based on the spirit's rambling monologue. The names Eddie and Charlie kept coming up and it was hard to tell who was who. But in fact, Elizabeth's husband was Edward Bullock, practically the first name out of the spirit/ Ethel's mouth. And Elizabeth had a brother— his name was Charles.

After that, most of what was learned by both normal and supernatural means turned out to be wrong. Elizabeth Bullock did not die as a result of her being hit by a car. The death certificate lists "chronic myocarditis," which is an inflammation of the muscle walls of the heart (but she did die in that drugstore). She also didn't speak with an Irish brogue. Elizabeth was born in New York, her father was German, and her mother, whose maiden name was Mary Schwieker, not Elizabeth McCuller, was born in the United States. Elizabeth and Edward didn't have any children that I could find, and New York does not have a death certificate for a child named Gracie Bullock.

What I was most curious about was: whatever happened to the tin of ashes? Slatin told me that the *Washington Post* had published a piece about his ghost in 1981, and a few months later the Slatins got a letter from a northern California priest named Devereaux. Father Devereaux offered to have a service for Elizabeth and to bury her in St. Patrick's Cemetery at Table Bluff, in Humboldt County, California. "Elizabeth will be resting with many of her own countrymen, in a very beautiful little cemetery," he wrote. Some of the tenants at Bank Street didn't want to see her go; she was a New York ghost after all.

I asked Slatin why they had never arranged for a Christian burial before this, since it was what the spirit had said she wanted. He told me that they'd taken Elizabeth's remains to a Catholic church in the city, but the church refused to bury her because she had married outside her faith. I hung up the phone, and then it hit me: Not only was this church being rather uncharitable, but if what Slatin had said were true, they were basing their decidedly uncharitable decision on information gained from a séance! This whole marrying-outside- ▶

When a Good Story Won't Let You Go
(continued)

her-faith thing came from a medium and was never confirmed outside the séance.

The Slatins decided that laying her to rest as she desired was the right thing to do, and so over their neighbors' objections they shipped her ashes to California. Fifty people attended the funeral mass at Father Devereaux's church in the town of Loleta. Like a movie, it poured the day they buried Elizabeth. Also, during the seance Betty had said, "I want a Christian burial in the shades of the Cross." And so they very kindly buried her beneath a cedar cross. The following Sunday, when Elizabeth was mentioned at mass, the lights in the church went out and it was so dark organist Doris Davey couldn't play.

How Elizabeth ended up in a ceiling at Bank Street when she was a resident of Perry Street long remained a mystery. But twenty-first-century technology is a wonderful resolution-provider. According to Edward Bullock's World War II draft registration card, now available online, Bullock moved out of their Perry Street apartment sometime after Elizabeth died and into smaller, more affordable accommodations in the rooming house at Bank Street, the very same rooming house the Slatins had converted. Why her tin of ashes were stored in the ceiling, I can't say. During our brief phone call Dr. Slatin, who doesn't believe in God or an afterlife, nonetheless admitted to me, "I felt her presence." The air in the apartment filled with cheap perfume whenever she made her appearance known.

I tried to learn more about Ethel Johnson Meyers, the medium who accompanied Hans Holzer to Bank Street. I didn't find out much, but as I say in the preface to this book, it always comes back to a love story, and this was true of Ethel Johnson Meyers.

Meyers, a former opera singer, was a trance medium and her control—the person who used her body to become a guide between the living and the dead—was her dead husband, a musician named Albert. Albert had died when his pharmacist made a mistake with his medicine and accidentally poisoned him. Ethel was going to walk into the sea and join him, but Albert came to her and stopped her. If she killed herself she wouldn't be with him, he warned. Just the opposite. "It would separate us." He told her that there was another way they could be together. Ethel went to a psychiatrist about the apparition, and he rather surprisingly suggested that she contact J. B. Rhine.

Soon afterward, Ethel found herself at the American Society for Psychical Research, which eventually led to her becoming a medium. Whenever she went into a trance, Albert appeared. However, it was a bittersweet and ultimately unsatisfying reunion for Ethel and Albert. They weren't really together again. Ethel was unconscious when Albert appeared, and when she awoke she remembered nothing. Hans Holzer spent more time with Albert than Ethel did. Even though Albert's spirit was inside her, somehow suffused throughout her body in order to communicate with whoever required his afterlife services, Albert was still as far apart from Ethel as he was the day he died.

I called Hans Holzer to ask him what Ethel and Albert were like, but he wasn't able to tell me anything. I had also intended to ask him if he ever tried to mediate any kind of exchange between his faithful medium and her beloved and deceased husband, e.g., "Hey, Albert, ▶

> ❝ How Elizabeth ended up in a ceiling at Bank Street when she was a resident of Perry Street remained a mystery. But twenty-first-century technology is a wonderful resolution-provider. ❞

When a Good Story Won't Let You Go
(continued)

as long as I've got you, any messages for Ethel?"
But this was not long before Holzer died and
he didn't seem to be feeling well, so I thanked
him and got off the phone more quickly than
I had planned. ❧

More about Research into the Paranormal

Something Hidden by Louisa E. Rhine,
McFarland & Company, 1983.

The Invisible Picture by Louisa E. Rhine,
McFarland Press, 1981.

There was so much I would have liked to have accomplished with this book, but to do what I originally intended would have taken a dozen books. Still, I regret that I didn't adequately portray the contribution of Dr. Louisa Rhine, whose role was at least as crucial as her husband's, if not more so. If he was the public face, I honestly think she was the private mind, the brains behind the operation.

The two most valuable books of hers to my research were *Something Hidden*, an intimate, moving account of her life with J. B. Rhine and their work, and *The Invisible Picture*. *The Invisible Picture* explains her theories about ghosts and was written for a general reader such as myself.

Parapsychology: An Insider's View of ESP by J. Gaither Pratt, Doubleday & Company, Inc., 1964.

Gaither Pratt: A Life for Parapsychology edited by Jurgen Keil, McFarland & Company, Inc., 1987.

My other regret is Gaither Pratt. He was the one who actually conducted much of the research the lab became famous for, and I wish he had loomed larger in my book as well.

Adventures in the Supernormal by Eileen J. Garrett, Helix Press, 2002.

More about Research into the Paranormal
(continued)

Mental Radio by Upton Sinclair, Hampton
Roads Publishing Company, Inc., 2001.
     I put these two books together because
I compared them side by side. Both try to
describe the process of what the authors
did, and were interesting peeks inside the
perhaps psychic process.

*The Elusive Science: Origins of Experimental
Psychical Research* by Seymour H. Mauskopf
and Michael R. McVaugh, The Johns Hopkins
University Press, 1980.
     An in-depth history of the science of
parapsychology. I couldn't have written my
book without it.

Parapsychology and the Skeptics by Chris Carter,
SterlingHouse Publisher, Inc., 2007.
     I started to write a chapter to address
why the Parapsychology Laboratory
experiments were never accepted by the
scientific community even though their
work appears sound. I quickly realized it
would need an entire book at least, and so
I abandoned the idea. Thankfully, Chris
Carter's book has since come out and it
addresses these issues very well.

*Extraordinary Knowing: Science, Skepticism,
and the Inexplicable Powers of the Human Mind*
by Elizabeth Lloyd Mayer, Ph.D., Bantam Books,
2007.
     At the end of my book I say that there seems
to be another source of information out there
that can't be explained. Dr. Mayer looks at this
and how science is addressing this mystery
today.

“ An in-depth
history of the
science of
parapsychology.
I couldn't have
written my book
without it. ”

Change the Rules! http://www.princeton.
edu/~pear/pdfs/Change_The_Rules.pdf
Robert G. Jahn and Brenda J. Dunne are
scientists whose work directly addresses the
additional source of information that they
describe as an information field. Their paper
discusses what science will need to do in order
to begin to understand it.

*Physics of the Impossible: A Scientific
Exploration into the World of Phasers,
Force Fields, Teleportation, and Time Travel*
by Michio Kaku, Doubelday, 2008.
   Dr. Kaku looks at things such as telepathy
and psychokinesis and discusses whether or
now they violate the laws of physics as we know
them and are therefore possible or impossible.

*The Gift: The Extraordinary Experiences of
Ordinary People* by Sally Rhine Feather and
Michael Schmicker, St. Martin's Paperbacks,
2006.
   The Rhine's daughter discusses her parent's
work and tells stories of ESP today.

*The Scalpel and the Soul: Encounters with
Surgery, the Supernatural, and the Healing Power
of Hope* by Allan J. Hamilton, MD, Tarcher,
2008.
   A memoir by a neurosurgeon whose stories
seem to confirm the Rhines's theories that
consciousness can exist independently of the
physical body.
*Entangled Minds: Extrasensory Experiences in a
Quantum Reality* by Dean Radin, Paraview
Pocket Books, 2006.
   A physicist explains how quantum physics
might help us to understand psi.

> " Dr. Kaku looks at things such as telepathy and psychokinesis and discusses whether or now they violate the laws of physics as we know them and are therefore possible or impossible. "

More about Research into the Paranormal
(continued)

The End of Materialism: How Evidence of the Paranormal is Bringing Science and Spirit Together by Charles T. Tart, Ph.D., New Harbinger Publications and the Institute of Noetic Sciences, 2009.

This is on my to-read pile. Charles Tart has been working on these questions since the fifties, and he both knew and sometimes locked horns with the Rhines. ∾